Mathematics: A First Course

Richard Meadows

Hutchinson

London Sydney Auckland Johannesburg

Hutchinson Education

An imprint of Century Hutchinson Ltd
62-65 Chandos Place, London WC2N 4NW

Century Hutchinson Australia Pty Ltd
89-91 Albion Street, Surry Hills
New South Wales 2010, Australia

Century Hutchinson New Zealand Limited
PO Box 40-086, Glenfield, Auckland 10,
New Zealand

Century Hutchinson South Africa (Pty) Ltd
PO Box 337, Bergvlei, 2012 South Africa

First published in 1989

Phototypesetting by Thomson Press (India) Limited New Delhi

Printed and bound in Great Britain by
Scotprint Ltd., Musselburgh

British Library Cataloguing in Publication Data
Meadows, R. G. (Richard Guy)
Mathematics
1. Mathematics
I. Title
510

ISBN 0-09-173006-6

Contents

Preface

This book is written for students taking a first course in mathematics. The contents follow exactly the learning objectives of the first level core unit of mathematics as published by the Business and Technician Education Council (BTEC).

Special emphasis has been placed in providing an easy-to-understand text backed at all stages by fully worked examples. Each chapter also contains a multiple-choice test and problems to aid revision and assessment.

The writing of this book has been a very interesting project and I sincerely hope students will find it helpful in strengthening their understanding and application of mathematics in solving practical problems.

Richard Meadows

January 1988

MATHEMATICS: A FIRST COURSE

Part One: Arithmetic

1 Indices and standard form

General learning objectives: to evaluate expressions involving indices and to use numbers expressed in standard form.

1.1 The power, index and base of a number

A number when successively multiplied by itself is said to be raised to a **power**. For example:

4×4 is four raised to the power 2 or 'four squared'
$5 \times 5 \times 5$ is five raised to the power 3 or 'five cubed'
$6 \times 6 \times 6 \times 6$ is six raised to the power 4

The mathematical way in which we express and write down the power of a number is

$4 \times 4 = 4^2$ ← index
↑ base number

$5 \times 5 \times 5 = 5^3$
$6 \times 6 \times 6 \times 6 = 6^4$
$7 \times 7 \times 7 \times 7 \times 7 = 7^5$

The number indicating the power to which the number is raised, that is the number of times the base number is multiplied by itself in the repeated multiplication, is known as the **index** (and also as the exponent). Indices (plural of index) are always indicated by the number which is raised and to the right of the **base** number.

A number is said to be expressed in indicial form when it is written as a base raised to a power. For example:

$100 = 10 \times 10 = 10^2$ in indicial form with 10 as base;
$32 = 2 \times 2 \times 2 \times 2 \times 2 = 2^5$ in indicial form and where in this case the base is 2 and the index is 5;
$81 = 9 \times 9 = 9^2$ with 9 as base, or alternatively with 3 as base;
$81 = 9 \times 9 = 3 \times 3 \times 3 \times 3 = 3^4$;
$10 = 10^1$ in indicial form, a trivial case really, the base being 10 and the index 1.

Examples

1 Write the following number in indicial form, i.e. as $(\text{base})^{\text{index}}$.
(a) 49 with 7 as base; (b) 27 with 3 as base; (c) 64 with 8 as base and also with 4 and 2 as bases

Solution

(a) $49 = 7 \times 7 = 7^2$
(b) $27 = 3 \times 9 = 3 \times 3 \times 3 = 3^3$
(c) $64 = 8 \times 8 = 8^2$ with 8 as base
$64 = 4 \times 4 \times 4 = 4^3$ with 4 as base
$64 = 2 \times 2 \times 2 \times 2 \times 2 \times 2 = 2^6$ with 2 as base

2 Evaluate (a) 8^3; $(\frac{1}{4})^2$; (c) 0.6^2

Solution

(a) $8^3 = 8 \times 8 \times 8 = 64 \times 8 = 512$

(b) $\left(\frac{1}{4}\right)^2 = \frac{1}{4} \times \frac{1}{4} = \frac{1}{16}$

(c) $0.6^2 = 0.6 \times 0.6 = 0.36$

3 Evaluate the following expressions:
(a) $3^2 + 4^2$; (b) $5^2 - 2^5$; (c) 6×3^2;
(d) $(12 \times 2^2) - (5 \times 2^3)$;
(e) $(3.6 \times 10^3) - (2.4 \times 10^3)$

Note: the power must always be evaluated first, then work out the multiplication and finally do the addition or subtraction.

Solution

(a) $3^2 + 4^2 = (3 \times 3) + (4 \times 4) = 9 + 16 = 25$

(b) $5^2 - 2^5 = (5 \times 5) - (2 \times 2 \times 2 \times 2 \times 2)$
$= 25 - 32 = -7$

(c) $6 \times 3^2 = 6 \times 9 = 54$

(d) $(12 \times 2^2) - (5 \times 2^3) = (12 \times 4) - (5 \times 8)$
$= 48 - 40 = 8$

(e) $(3.6 \times 10^3) - (2.4 \times 10^3) = (3.6 \times 1000)$
$- (2.4 \times 1000)$
$= 3600 - 2400 = 1200$

An extra note on base

The term base is used also in a very fundamental way in mathematics to define the numerical system we are using. The most common system of numbers and the ones we use in everyday work is the denary or decimal system. These are base 10 numbers. We write numbers based on powers of 10. For example:

5 7 2 4
- 4 → number of units, 4
- 2 → number of 10s, 2×10^1
- 7 → number of 100s, 7×10^2
- 5 → number of 1000s, 5×10^3

As we move from right to left in a denary (base 10) number the magnitude of the digit is increased by a power of 10.

The simplest of all number systems is the binary system which only has two digits, 0 and 1. The binary system is a base 2 system and as we move from right to left in a binary number the magnitude of the binary digit (usually abbreviated to bit) increases by a power of 2. For example, the binary number 101 is equal to the decimal or denary base 10 number of 5:

1 1 1
- 1 → 1 unit
- 1 → 0×2^1 (zero twos)
- 1 → 1×2^2

i.e. $111_2 = 1 \text{ (unit)} + (0 \times 2^1) + (1 \times 2^2)$
$= 1 + 0 + 4 = 5_{10}$,
so binary 111 = denary 5.

To distinguish number systems we can add a subscript at the lower right of the number to indicate the base – we have done this above.

1.2 The reciprocal of a number

The **reciprocal** of a number is *one divided by the number*. The product of a number and its reciprocal equals one (unity).
For example:

the reciprocal of $2 = \frac{1}{2} = 0.5$

the reciprocal of $3 = \frac{1}{3} = 0.3333\cdots$ (0.3 recurring)

the reciprocal of $\frac{1}{4} = 1 \div \frac{1}{4} = 4$

the reciprocal of $0.2 = \frac{1}{0.2} = 5$

We write reciprocals in indicial form using -1 as the index, i.e.

$a^{-1} = \frac{1}{a} \cdots$ the reciprocal of a

so 10^{-1} means $\frac{1}{10} = 0.1$

100^{-1} means $\frac{1}{100} = 0.01$

0.25^{-1} means $\frac{1}{0.25} = 4$

In general the meaning of a negative value index is given by:

$$a^{-n} = \frac{1}{a^n}$$

So, for example

$$10^{-2} = \frac{1}{10^2} = \frac{1}{100} = 0.01$$

$$10^{-3} = \frac{1}{10^3} = \frac{1}{1000} = 0.001$$

$$2^{-4} = \frac{1}{2^4} = \frac{1}{16}$$

Examples

1 Determine the reciprocals of
(a) 5 (b) 8 (c) 20

Solution

(a) $5^{-1} = \frac{1}{5} = 0.2$

(b) $8^{-1} = \frac{1}{8} = 0.125$

(c) $20^{-1} = \frac{1}{20} = 0.05$

Remember: To convert a fraction into a decimal divide the denominator into the numerator, so

$$\frac{1}{5} = \begin{array}{r} 0.2 \\ 5\overline{)1.0} \end{array} \qquad \frac{1}{8} = \begin{array}{r} 0.125 \\ 8\overline{)1.000} \\ \underline{8} \\ 20 \\ \underline{16} \\ 40 \\ \underline{40} \\ .. \end{array} \qquad \frac{1}{20} = \begin{array}{r} 0.05 \\ 20\overline{)1.00} \\ \underline{100} \\ \ldots \end{array}$$

2 Determine the reciprocals of

(a) $\frac{2}{3}$ (b) 0.4 (c) 12.5

Solution

(a) $\left(\frac{2}{3}\right)^{-1} = 1 \div \frac{2}{3} = 1 \times \frac{3}{2} = \frac{3}{2} = 1\frac{1}{2}$ or 1.5

(b) $0.4^{-1} = \frac{1}{0.4} = \frac{1}{0.4} \times \frac{10}{10} = \frac{10}{4} = 2\frac{1}{2}$ or 2.5

(c) $12.5^{-1} = \frac{1}{12.5} = \begin{array}{r} 0.08 \\ 125\overline{)10.00} \\ \underline{1000} \\ \ldots. \end{array}$,

i.e. $12.5^{-1} = 0.08$

3 Evaluate $\left(\frac{1}{60} + \frac{1}{30}\right)^{-1}$

Solution

First work out the addition within the brackets,

$$\frac{1}{60} + \frac{1}{30} = \frac{1+2}{60} = \frac{3}{60} = \frac{1}{20}$$

then find the reciprocal,

$$\left(\frac{1}{60} + \frac{1}{30}\right)^{-1} = \left(\frac{1}{20}\right)^{-1} = 20$$

1.3 Rules of indices

1.3.1 Multiplication rule

$$a^m \times a^n = a^{m+n}$$

i.e. *Add* indices to find the product of two or more numbers, provided these numbers have the same base

For example,

$$2^2 \times 2^3 = 2^{2+3} = 2^5$$

Check $2^2 = 2 \times 2 = 4$

$2^3 = 2 \times 2 \times 2 = 8$

so $2^2 \times 2^3 = 4 \times 8 = 32$

but $2^5 = 2 \times 2 \times 2 \times 2 \times 2 = 32$

Examples

1 Evaluate (a) $5^3 \times 5^4$ (b) $1.2^2 \times 1.2^3$

Solution

(a) $5^3 \times 5^4 = 5^{3+4} = 5^7$ $(= 78\,125)$

(b) $1.2^2 \times 1.2^3 = 1.2^5$ $(= 2.48832)$

2 Evaluate (a) $(2 \times 10^3) \times (4 \times 10^2)$
(b) $5 \times 10^6 \times 3 \times 10^{-3}$

Solution

Note: whenever we have numbers 'in front of' indicial forms we must first work out their product.

(a) $(2 \times 10^3) \times (4 \times 10^2) = 2 \times 4 \times 10^{3+2}$
$= 8 \times 10^5$

(b) $5 \times 10^6 \times 3 \times 10^{-3} = 5 \times 3 \times 10^{6+(-3)}$
$= 15 \times 10^{6-3}$
$= 15 \times 10^3$

1.3.2 Division rule

$$a^m \div a^n = \frac{a^m}{a^n} = a^{m-n}$$

i.e. *Subtract* indices to carry out the division of two numbers expressed in indicial form with the same base.
For example:

$$3^5 \div 3^2 = \frac{3^5}{3^2} = 3^{5-2} = 3^3$$

Check $3^5 = 3 \times 3 \times 3 \times 3 \times 3 = 243$
$3^2 = 3 \times 3 = 9$
so $3^5 \div 3^2 = 243 \div 9 = 27$
but $3^3 = 27$

Examples

1 Evaluate (a) $5^{14} \div 5^9$ (b) $6.3^5 \div 6.3^2$

Solution

(a) $5^{14} \div 5^9 = 5^{14-9} = 5^5$
(b) $6.3^5 \div 6.3^2 = 6.3^{5-2} = 6.3^3$

2 Evaluate (a) $(9.9 \times 10^6) \div (3.3 \times 10^4)$
(b) $6.4 \times 10^5 \times 9.8 \times 10^3 \div 3.2 \times 10^6$

Solution

(a) $(9.9 \times 10^6) \div (3.3 \times 10^4)$

$$= \frac{9.9 \times 10^6}{3.3 \times 10^4}$$

$$= \frac{9.9}{3.3} \times 10^{6-4} = 3.0 \times 10^2$$

(b) $6.4 \times 10^5 \times 9.8 \times 10^3 \div 3.2 \times 10^6$

$$= \frac{6.4 \times 9.8}{3.2} \times 10^{5+3-6}$$

$$= \frac{\overset{2}{\cancel{6.4}} \times 9.8}{\underset{1}{\cancel{3.2}}} \times 10^2$$

$$= 19.6 \times 10^2 = 1960$$

1.3.3 $a^0 = 1$

Any number raised to the power 0 equals 1.
For example,

$$4^0 = 1 \qquad 10^0 = 1 \qquad 88.97^0 = 1$$

This result can be deduced from the division rule:

$a^m \div a^m$ obviously equals 1
but $a^m \div a^m = a^{m-m} = a^0$
so $a^0 = 1$

1.3.4 $a^{-n} = 1/a^n$

This result can also be deduced from the division rule and using the fact that $a^0 = 1$:

$$1 \div a^n = \frac{1}{a^n} = \frac{a^0}{a^n} = a^{0-n} = a^{-n}$$

Examples

1 Evaluate (a) 2^{-3} (b) 5^{-2}

Solution

(a) $2^{-3} = \frac{1}{2^3} = \frac{1}{8}$ (b) $5^{-2} = \frac{1}{5^2} = \frac{1}{25}$

2 Evaluate (a) $3.6 \times 10^3 \div 1.8 \times 10^6$
(b) $4.2 \times 10^{-3} \times 2.0 \times 10^{-6}$

Solution

(a) $3.6 \times 10^3 \div 1.8 \times 10^6$

$$= \frac{3.6 \times 10^3}{1.8 \times 10^6}$$

$$= \frac{3.6}{1.8} \times 10^{3-6} = 2.0 \times 10^{-3}$$

$$= \frac{2.0}{10^3} = 0.002$$

(b) $4.2 \times 10^{-3} \times 2.0 \times 10^{-6}$
$= 4.2 \times 2 \times 10^{-3+(-6)}$
$= 8.4 \times 10^{-3-6} = 8.4 \times 10^{-9}$

1.3.5 Raising to a power: $(a^m)^n = a^{mn}$

To find the power of a number expressed in indicial form, *multiply* the indices.
For example:

$$(2^2)^3 = 2^{2 \times 3} = 2^6$$

Check $2^2 = 4$
$4^3 = 4 \times 4 \times 4 = 64$
but $2^6 = 2 \times 2 \times 2 \times 2 \times 2 \times 2 = 64$

Examples

1 Evaluate (a) $(3^3)^2$ (b) $(10^3)^6$

Solution

(a) $(3^3)^2 = 3^{3\times 2} = 3^6$
(b) $(10^3)^6 = 10^{3\times 6} = 10^{18}$

2 Evaluate (a) $(5 \times 10^3)^2$ (b) $(1.2 \times 10^{-3})^2$

Solution

(a) $(5 \times 10^3)^2 = 5^2 \times 10^{3\times 2} = 25 \times 10^6$
(b) $(1.2 \times 10^{-3})^2 = 1.2^2 \times 10^{-3\times 2} = 1.44 \times 10^{-6}$

Note: all terms within the brackets must be raised to the power, i.e. in the case of $(5 \times 10^3)^2$ we must work out 5^2 as well as use the rule of indices to work out $(10^3)^2$.

1.3.6 Summary of rules of indices

Multiplication: $a^m \times a^n = a^{m+n}$
Division: $a^m / a^n = a^{m-n}$
Raising to a power: $(a^m)^n = a^{mn}$

$a^0 = 1, \quad a^{-1} = \frac{1}{a}, \quad a^{-n} = \frac{1}{a^n}$

1.4 Standard form

In science, engineering, business and indeed in everyday life we frequently encounter a very wide range of numerical values from the very large to the very small.
For example:

The velocity of light is

300 000 000 metres per second
or 671 000 000 miles per hour;

the distance of the planet Pluto from the sun is

5 900 000 000 kilometres;

the debt of a country may be

$ 10 000 000;

whilst,
the amount of the rare gas xenon in the earth's atmosphere is

0.000 01%

and the charge on an electron is

−0.000 000 000 000 000 000 016 coulombs

and the wavelengths of light are in the approximate range

0.000 000 000 8 down to 0.000 000 000 4 metres;

Clearly such very large and exceedingly small numbers are very cumbersome both to comprehend and write down. To overcome this difficulty and also to aid calculations it is common practice to express numbers in standard form.

The **standard form** of a number consists of expressing the number in decimal form with only one digit in front of the decimal point and multiplied by the correct power of 10 to give the required number, i.e.

standard form $= A \times 10^n$

where A is between 1.0000... and 9.9999... and the index n is the required power of 10.
For example:

42.3 in standard form is 4.23×10^1
598 in standard form is 5.98×10^2
9999.7 in standard form is 9.9997×10^3
0.67 in standard form is 6.7×10^{-1}
0.018 in standard form is 1.8×10^{-2}
0.00757 in standard form is 7.57×10^{-3}

and the quantities we used to introduce the section are, in standard form, expressed as

velocity of light $= 3 \times 10^8$ m/s
or 6.71×10^8 mile/h
distance of Pluto $= 5.9 \times 10^9$ km
debt $= \$1.0 \times 10^7$
electronic charge $= 1.6 \times 10^{-9}$ C
wavelength range of light $= 8.0 \times 10^{-10}$ to 4.0×10^{-10} m

Rules for converting a number to standard form
Count the number of decimal places required to move the decimal point to convert the number to 'one digit before the decimal point'; the number of places moved gives the index for the power of 10 multiplier. The index is positive if the decimal point is moved from right to left (←) and negative if moved from left to right (→).
For example:

36 000 000; so $36\,000\,000 = 3.6 \times 10^7$
7 places

0000 00 269 so $0.000\,002\,69 = 2.69 \times 10^{-6}$
6 places

To convert a number expressed in standard form back to 'normal' form
The index tells us the number of places to move the decimal point:

if it is positive move the decimal point left to right (→);

if it is negative move the decimal point right to left (←)

writing-in zeros for all places moved after the first step.
For example:

$6.97 \times 10^{-3} = .00697,$ (3 places) so $6.97 \times 10^{-3} = 0.006\,97$

$9.126 \times 10^{6} = 9\,126\,000.$ (6 places) so $9.126 \times 10^{6} = 9\,126\,000$

Examples

1 Write the following numbers in standard form:
(a) 2266 (b) 0.01499 (c) 567.234

Solution

(a) $2266 = 2.266 \times 10^{3}$
(b) $0.01499 = 1.499 \times 10^{-2}$
(c) $567.234 = 5.672\,34 \times 10^{2}$

2 Convert the following from standard form to 'normal' numbers:
(a) 9.3×10^{6} (b) 4.6×10^{-3} (c) 1.8×10^{-6}

Solution

(a) $9.3 \times 10^{6} = 9\,300\,000$
(b) $4.6 \times 10^{-3} = 0.0046$
(c) $1.8 \times 10^{-6} = 0.000\,001\,8$

1.5 Calculations using standard form

Calculations can often best be made by first expressing numbers in standard form, especially so if we want to make an approximate calculation. There are, however, some important points to note when handling numbers in standard form.

Addition and subtraction

When adding or subtracting numbers in standard form we must always make sure the numbers are first arranged to have the same index (power of 10) or we must convert both back to normal decimal form.

Examples

1 Evaluate $(7.4 \times 10^{3}) + (2.6 \times 10^{4})$

Solution

Converting 7.4×10^{3} to a number the same power 4 as 2.6×10^{4} we have:

$$7.4 \times 10^{3} = 0.74 \times 10^{4}$$

so $(7.4 \times 10^{3}) + (2.6 \times 10^{4}) = (0.74 \times 10^{4}) + (2.6 \times 10^{4}) = (0.74 + 2.6) \times 10^{4} = 3.34 \times 10^{4}$

Alternatively, converting the numbers to normal form:

$7.4 \times 10^{3} = 7400$
$2.6 \times 10^{4} = 26\,000$

Adding $= 33\,400$

so $(7.4 \times 10^{3}) + (2.6 \times 10^{4}) = 33\,400 = 3.34 \times 10^{4}$

2 Evaluate $(5.7 \times 10^{-2}) - (2.9 \times 10^{-3})$

Solution

$5.7 \times 10^{-2} = 0.0570$
$2.9 \times 10^{-3} = 0.0029$

Subtracting $= 0.0541$

Alternatively:

$(5.7 \times 10^{-2}) - (2.9 \times 10^{-3})$
$= (5.7 \times 10^{-2}) - (0.29 \times 10^{-2})$
$= (5.7 - 0.29) \times 10^{-2} = 5.41 \times 10^{-2}$

Multiplication and division

Here we multiply/divide the decimal number part of the standard form separately and then use the rule of indices to deal with power of 10 parts.

Examples

1 Evaluate (a) $(3 \times 10^{6}) \times (6 \times 10^{4})$
(b) $(8 \times 10^{4}) \div (4 \times 10^{5})$

Solution

(a) $(3 \times 10^6) \times (6 \times 10^4)$
$= 3 \times 6 \times 10^{6+4}$
$= 18 \times 10^{10} = 1.8 \times 10^{11}$

(b) $\frac{8 \times 10^4}{4 \times 10^5} = \frac{8}{4} \times 10^{4-5} = 2 \times 10^{-1}$ (or 0.2)

2 Evaluate $\frac{(5.25 \times 10^6) - (1.25 \times 10^6)}{(2.42 \times 10^3) - (1.42 \times 10^3)}$

Solution

$$\frac{(5.25 \times 10^6) - (1.25 \times 10^6)}{(2.42 \times 10^3) - (1.42 \times 10^3)}$$
$$= \frac{(5.25 - 1.25)10^6}{(2.42 - 1.42)10^3}$$
$$= \frac{4.0 \times 10^6}{1.0 \times 10^3} = 4.0 \times 10^{6-3} = 4.0 \times 10^3$$

Test and problems 1

Multiple choice test: MT 1

Answer block:

Question No.	0	1	2	3	4	5	6	7	8
Answer	c								

Enter your answer, that is a, b, c or d in the column under the question number in the answer block above.
Note that question Qu. 0 has already been worked out and the answer inserted.

Qu. 0 Evaluate $(8 \times 10^3) \times (2 \times 10^5)$ and express the result in standard form.
Ans (a) 16×10^8 (b) 10^9
(c) 1.6×10^9 (d) 1.6×10^2

Solution

$(8 \times 10^3) \times (2 \times 10^5) = (8 \times 2) \times 10^{3+5}$
$= 16 \times 10^8 = 1.6 \times 10^9$

so the correct answer is (c) and we insert 'c' under Qu. No. 0 in the answer block.

Now carry on with the test.

Qu. 1 Evaluate $4^2 + 2^3$
Ans (a) 14 (b) 22 (c) 10 (d) 24

Qu. 2 Express 81 in indicial form with 3 as base
Ans (a) 9 (b) 3^4 (c) 3 (d) 8.1×10^1

Qu. 3 Determine the reciprocal of 25
Ans (a) 1 (b) 5 (c) $\frac{1}{25}$ (d) 0.05

Qu. 4 Evaluate $3^2 \times 2^3$
Ans (a) 36 (b) 3^5 (c) 30 (d) 72

Qu. 5 Evaluate $6^5 \div 6^3$
Ans (a) 5 (b) 36 (c) 6^8 (d) 25

Qu. 6 Evaluate $(3^2)^3$
Ans (a) 729 (b) 81 (c) 243 (d) 27

Qu. 7 Express 0.000 006 7 in standard form
Ans (a) 6.7×10^6 (b) 6.7×10^{-6}
(c) 6.7 (d) 6.7×10^{-5}

Qu. 8 Evaluate $\frac{(2 \times 10^6) + (4 \times 10^6)}{(5 \times 10^3) - (4 \times 10^3)}$
Ans (a) 6×10^6 (b) 6×10^3
(c) 600 (d) 6.7×10^2

Problems 1

1 Express the following numbers in indicial form:
(a) 121 with 11 as base
(b) 512 with 2 as base
(c) 10 000 with 10 as base
(d) 125 with 5 as base

2 Determine the reciprocals of:
(a) 4 (b) 100 (c) 0.2 (d) $\frac{7}{8}$

3 Evaluate the following:
(a) 2^4 (b) 4^2 (c) $\left(\frac{1}{3}\right)^2$ (d) 1.2^2

4 Determine the values of the following products:
(a) $4^2 \times 4^2$ (b) $5^2 \times 2^5$ (c) $4 \times 2^3 \times 3 \times 2^2$

5 Evaluate:
(a) $10^6 \div 10^2$ (b) $(4.2 \times 10^3) \div (7.0 \times 10^2)$
(c) $7^7 \div 7^5$ (d) $2^{10} \times 2^8 \div 2^{12}$

6 Express the following numbers in standard form:
(a) 24 000 (b) 0.0024 (c) $\frac{1}{2}$ (d) 999.9

7 Evaluate, expressing the result in standard form:

(a) $(6.2 \times 10^6) + (3.1 \times 10^6) - (9.3 \times 10^5)$

(b) $(3.0 \times 10^{-3}) + (6.0 \times 10^{-4})$

(c) $(5.2 \times 10^{-2}) - (2.6 \times 10^{-3})$

(d) $(6 \times 10^3)^2$

(e) $4.0 \times 10^6 \times 2.0 \times 10^3 \times 3.0 \times 10^{-6}$

(f) $(6.6 \times 10^6) \div (3.3 \times 10^9)$

2 Indices and logarithms

General learning objectives: to evaluate expressions involving negative and fractional indices and to relate indices logarithms.

2.1 Positive, negative and fractional indices: powers and roots

In the first chapter we dealt with whole number indices. We explained that when the index n is a positive integer, i.e. when $n = 1, 2, 3, 4 \ldots$ etc.,

$$a^n = a \times a \times a \cdots \times a$$
(a multiplied by itself n times)

when $n = 0$, we had the special case,

$$a^0 = 1$$

any number raised to the power 0 equals one:

e.g. $9.9^0 = 1, \quad 100^0 = 1, \quad 0.03^0 = 1$

We also defined the reciprocal of a number:

the reciprocal of a, denoted by $a^{-1} = \dfrac{1}{a}$

e.g. $3^{-1} = \dfrac{1}{3}, \quad 10^{-1} = \dfrac{1}{10}, \quad 0.4^{-1} = \dfrac{1}{0.4} = 2.5$

and also introduced the concept of the meaning of a negative index as,

$$a^{-n} = \frac{1}{a^n}$$

e.g. $2^{-3} = \dfrac{1}{2^3} = \dfrac{1}{2 \times 2 \times 2} = \dfrac{1}{8}$

We then proceeded to define the laws of indices and apply them to evaluate expressions involving whole number indices. The laws for fractional indices are exactly the same but before using them for fractional case we must first consider the meaning of a fractional index. Let us first consider a simple fractional index and the concept of the 'root' of a number.

The square root of a number is either one of two equal numbers which when multiplied by itself gives the number. The symbol used to denote the square root is the $\sqrt{\ }$ symbol. For example, the square root of 4 is written as $\sqrt{4}$ and since,

$$4 = 2 \times 2$$
$$\sqrt{4} = 2$$

Similarly, as $9 = 3 \times 3, \quad \sqrt{9} = 3$
as $16 = 4 \times 4, \quad \sqrt{16} = 4$
as $25 = 5 \times 5, \quad \sqrt{25} = 5$
as $144 = 12 \times 12, \quad \sqrt{144} = 12$

Such numbers are termed perfect squares as they have exact whole number square roots. In general, however, the square root of a number will not be so simple. It usually has a decimal part as well. Square roots can easily be found using a calculator or by looking up square root tables as we explain in Chapter 4. For example,

$\sqrt{12.96} = 3.6, \quad 3.6 \times 3.6 = 12.96$
$\sqrt{0.84} = 0.9165, \quad 0.9165 \times 0.9165 = 0.84$

It is also important to note that the square root of a number may be either positive or negative. For example,

$$3 \times 3 = 9 \quad \text{so } \sqrt{9} = 3$$

but $-3 \times -3 = 9$, hence $\sqrt{9}$ is also given -3, since when we multiply two negatives together we obtain a positive. The fact that the square root may be positive or negative can be expressed using the $\pm$ symbol, e.g.

$\sqrt{9} = \pm 3$, meaning the square root of 9 has two values: $+3$ and -3.

In indicial form we denote the square root of a number using the fractional index $\dfrac{1}{2}$, i.e. the square root of a is written as

$$\sqrt{a} = a^{\frac{1}{2}}$$

The cube root of a number is one of three equal numbers which when multiplied together gives the number. The cube root is denoted by the symbol $\sqrt[3]{\ }$ and in indicial form by the fractional index $\frac{1}{3}$, i.e.

$$\sqrt[3]{a} = a^{\frac{1}{3}}$$

So, for example,

$$\sqrt[3]{8} = 8^{\frac{1}{3}} = 2 \qquad (\text{as } 2 \times 2 \times 2 = 8)$$
$$\sqrt[3]{27} = 27^{\frac{1}{3}} = 3 \qquad (\text{as } 3 \times 3 \times 3 = 27)$$
$$\sqrt[3]{125} = 125^{\frac{1}{3}} = 5 \qquad (\text{as } 5 \times 5 \times 5 = 125)$$

The cube root of a positive number is always positive. Although there can be no 'real' square root of a negative number, a negative number does have a cube root and this is negative, e.g.

$$\sqrt[3]{(-64)} = -4, \quad \text{as } -4 \times -4 \times -4 = -64$$

The fourth root of a number is one of four equal numbers which when multiplied together gives the number. The fourth root is denoted in indicial form using the fractional index $\frac{1}{4}$, i.e.

$$\sqrt[4]{a} = a^{\frac{1}{4}}$$

e.g. $\sqrt[4]{81} = 81^{\frac{1}{4}} = \pm 3 \quad \text{as } 3 \times 3 \times 3 \times 3 = 81$
and as $\quad -3 \times -3 \times -3 \times -3 = 81$

Likewise the fifth root is one of five equal numbers whose product equals the number, the sixth root is one of six equal numbers which when multiplied together gives the number... and so on. In indicial form:

$$a^{\frac{1}{5}} = \text{fifth root of } a, \quad a^{\frac{1}{6}} = \text{sixth root of } a$$

For example, $32^{\frac{1}{5}} = 2$
$729^{\frac{1}{6}} = \pm 3$

In all the above examples we selected numbers which happened to have exact whole number roots. In general, as we noted with square roots, the root of a number will not normally be an exact integer (whole number). Roots can be evaluated on most electronic calculators.

Now let us move on to consider the meaning of a fractional index $n = p/q$ where p and q are whole numbers.

$a^{p/q}$ means the qth root of a raised to the power p

For example,

$$16^{\frac{3}{4}} = (16^{\frac{1}{4}})^3 = (2)^3 = 8$$

i.e. the 4th root of 16, which is 2, raised to the power 3;

$$27^{\frac{2}{3}} = (27^{\frac{1}{3}})^2 = (3)^2 = 9$$
$$121^{\frac{3}{2}} = (121^{\frac{1}{2}})^3 = 11^3 = 11 \times 11 \times 11 = 1331$$

We can also have negative fractional indices. Remember a negative index means:

$$a^{-n} = \frac{1}{a^n}$$

so if $n = -p/q$,

$$a^{-p/q} = \frac{1}{a^{p/q}}$$

$a^{-p/q}$ is simply one divided by $a^{p/q}$, the reciprocal of $a^{p/q}$. For example:

$$16^{-\frac{3}{2}} = \frac{1}{16^{\frac{3}{2}}} = \frac{1}{(16^{\frac{1}{2}})^3} = \frac{1}{4^3} = \frac{1}{64}$$

$$27^{-\frac{1}{3}} = \frac{1}{27^{\frac{1}{3}}} = \frac{1}{3}$$

$$49^{-\frac{5}{2}} = \frac{1}{49^{\frac{5}{2}}} = \frac{1}{(49^{\frac{1}{2}})^5} = \frac{1}{7^5} = \frac{1}{16807}$$

2.2 Index rules for negative and fractional indices

The laws of indices are quite general and may be applied for positive, negative and fractional indices in exactly the same way as previously considered in section 1.3. Thus whether the indices m and n are positive or negative, whole number or fractional, we have:

Multiplication $a^m \times a^n = a^{m+n}$ ... *to multiply, add indices*

e.g. $2^2 \times 2^3 = 2^{2+3} = 2^5 = 32$

$$2^{\frac{1}{2}} \times 2^{\frac{3}{2}} = 2^{\frac{1}{2}+\frac{3}{2}} = 2^2 = 4$$

$$2^{-1} \times 2^{-3} = 2^{-1+(-3)} = 2^{-4} = \frac{1}{16}$$

$$2^{-\frac{1}{2}} \times 2^{+\frac{5}{2}} = 2^{-\frac{1}{2}+\frac{5}{2}} = 2^2 = 4$$

Division $\dfrac{a^m}{a^n} = a^{m-n}$ ... *to divide, subtract indices* e.g. $\dfrac{3^7}{3^4} = 3^{7-4} = 3^3 = 27$

$\dfrac{10^{\frac{3}{2}}}{10^{\frac{1}{2}}} = 10^{\frac{3}{2}-\frac{1}{2}} = 10^1 = 10$

$\dfrac{81^{\frac{1}{2}}}{81^{\frac{1}{4}}} = 81^{\frac{1}{2}-\frac{1}{4}} = 81^{\frac{1}{4}} = \pm 3$

$\dfrac{10^{-3}}{10^{-5}} = 10^{-3-(-5)} = 10^{-3+5} = 10^2$

Raising to a power $(a^m)^n = a^{m \times n}$ ... to raise to a power, multiply indices

e.g. $(10^3)^4 = 10^{3\times 4} = 10^{12}$

$(10^{-3})^2 = 10^{-3\times 2} = 10^{-6}$

$(4^{\frac{1}{2}})^3 = 4^{\frac{3}{2}} = 2^3 = 8$

$(25^{\frac{1}{2}})^{-2} = 25^{\frac{1}{2}\times -2} = 25^{-1} = \dfrac{1}{25}$

Examples

1 Evaluate $(4^{-\frac{1}{2}}) + 2^{-2}$

Solution

$4^{-\frac{1}{2}} = \dfrac{1}{4^{\frac{1}{2}}} = \dfrac{1}{2}, \quad 2^{-2} = \dfrac{1}{2^2} = \dfrac{1}{4}$

so $4^{-\frac{1}{2}} + 2^{-2} = \dfrac{1}{2} + \dfrac{1}{4} = \dfrac{3}{4}$

2 Evaluate:
(a) $2^{\frac{1}{3}} \times 2^{\frac{1}{2}} \times 2^{\frac{1}{6}}$; (b) $3^{\frac{1}{2}} \div 3^{3\frac{1}{2}}$; (c) $10^3 \div 10^{2.5}$

Solution

(a) $2^{\frac{1}{3}} \times 2^{\frac{1}{2}} \times 2^{\frac{1}{6}} = 2^{(\frac{1}{3}+\frac{1}{2}+\frac{1}{6})} = 2^1 = 2,$

as $\dfrac{1}{3} + \dfrac{1}{2} + \dfrac{1}{6} = \dfrac{2+3+1}{6} = \dfrac{6}{6} = 1$

(b) $3^{\frac{1}{2}} \div 3^{3\frac{1}{2}} = \dfrac{3^{\frac{1}{2}}}{3^{3\frac{1}{2}}} = 3^{\frac{1}{2}-3\frac{1}{2}} = 3^{-3} = \dfrac{1}{3^3} = \dfrac{1}{27}$

(c) $10^3 \div 10^{2.5} = 10^{3-2.5} = 10^{\frac{1}{2}} = 3.1623$
(using calculator)

3 Evaluate $(0.1)^2 + (0.1)^{-2} + (1.44)^{\frac{1}{2}}$

Solution

$0.1^2 = 0.1 \times 0.1 = 0.01$

$0.1^{-2} = \dfrac{1}{0.1^2} = \dfrac{1}{0.01} = 100$

$1.44^{\frac{1}{2}} = 1.2$, as $1.2 \times 1.2 = 1.44$
(taking positive root).

So $0.1^2 + 0.1^{-2} + 1.44^{\frac{1}{2}}$
$= 0.01 + 100 + 1.2 = 101.21$

4 Evaluate (a) $5^{\frac{1}{2}} \times 5^{-\frac{1}{2}}$; (b) $10^{3.2} \div 10^{1.2}$;
(c) $2^{-1.6} \times 2^{0.6} \div 2^3$

Solution

(a) $5^{\frac{1}{2}} \times 5^{-\frac{1}{2}} = 5^{\frac{1}{2}-\frac{1}{2}} = 5^0 = 1$

(b) $10^{3.2} \div 10^{1.2} = 10^{3.2-1.2} = 10^2 = 100$

(c) $2^{-1.6} \times 2^{0.6} \div 2^3$

$= 2^{-1.6+0.6-3} = 2^{-4} = \dfrac{1}{2^4} = \dfrac{1}{16}$

2.3 The definition of the logarithm of a number: the inverse of $a^x = y$ as $x = \log_a y$

Closely associated with indices is the logarithm of a number.

If a positive number y is expressed in indicial form with base a, i.e.

$y = a^x$

then the index x is known as the logarithm of y to base a. Writing this in mathematic terms, we have:

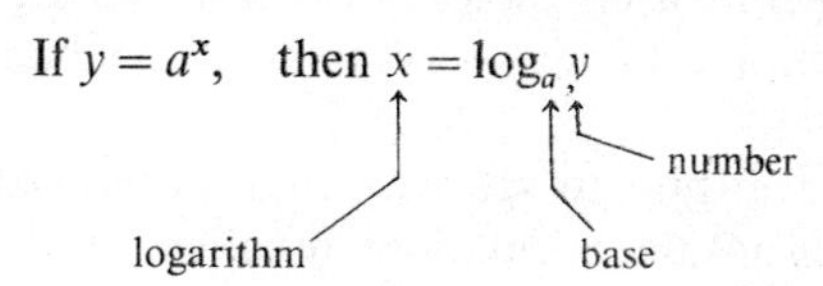

For example:

1 If 32 is expressed as $32 = 2^5$ then the logarithm of 32 to base 2 is 5,
i.e. if $y = 32 = 2^5$, then $\log_2 32 = 5$

2 If 16 is expressed as $16 = 4^2$, then $\log_4 16 = 2$

3 If 81 is expressed as $81 = 3^4$, then $\log_3 81 = 4$

4 If $\log_5 y = 2$, then $y = 5^2 = 25$

5 If $\log_{10} y = 3$, then $y = 10^3 = 1000$

6 If $\log_{10} y = -3$, then $y = 10^{-3} = \dfrac{1}{1000}$

2.4 Common logarithms: logs to base 10

The most commonly used form of logarithms (logs for short) are to base 10. They are used widely in science and engineering and before the widespread use of electronic calculators they were an important aid to working-out calculations.

Logs to base 10 are known as common logarithms and using the general definition:

$$y = a^x, \quad x = \log_a y$$

with 10 as base, we have for the common logarithm of any positive number N:

$$N = 10^x, \quad \text{then } \log_{10} N = x$$

It is an easy matter to deduce the common log of numbers that can be expressed as whole number power of 10, simply by applying the above definition:

$$\begin{aligned} 1000 &= 10^3 & \text{so } \log_{10} 1000 &= 3 \\ 100 &= 10^2 & \text{so } \log_{10} 100 &= 2 \\ 10 &= 10^1 & \text{so } \log_{10} 10 &= 1 \\ 1 &= 10^0 & \text{so } \log_{10} 1 &= 0 \\ 0.1 &= 10^{-1} & \text{so } \log_{10} 0.1 &= -1 \\ 0.01 &= 10^{-2} & \text{so } \log_{10} 0.01 &= -2 \\ 0.001 &= 10^{-3} & \text{so } \log_{10} 0.001 &= -3 \end{aligned}$$

Note that the logs of 0.1, 0.01, 0.001,... have negative values.

For all other numbers, the common log can be found either by using tables or by using the **log** key on an electronic calculator as explained in Chapters 4 and 5.

It is interesting to see why logs were used so extensively to aid calculations involving multiplication, division and finding powers and roots. The reason follows by reminding ourselves of the laws of indices:

let N_1 and N_2 be numbers expressed in indicial form to base 10, i.e. suppose

$$N_1 = 10^{x_1} \quad \text{and} \quad N_2 = 10^{x_2}$$

then $\log_{10} N_1 = x_1$ and $\log_{10} N_2 = x_2$

1 *Multiplication: add indices (add logs)*

$$N_1 \times N_2 = 10^{x_1} \times 10^{x_2} = 10^{x_1 + x_2}$$

so $\log_{10} N_1 \times N_2 = x_1 + x_2$
i.e. to find the log of a product, **add logs**.

2 *Division: subtract indices (subtract logs)*

$$\frac{N_1}{N_2} = \frac{10^{x_1}}{10^{x_2}} = 10^{x_1 - x_2}$$

so $\log_{10} N_1/N_2 = x_1 - x_2$
i.e. to find the log of a division, **subtract logs**.

Thus we see by using logs we have essentially converted multiplication into an addition process and division into a subtraction process. Addition and subtraction are, of course, easier processes than multiplication and division. Tables had to be used: log tables to look up the logs of numbers and antilog tables to reverse the process to find the number corresponding to the 'log result'.

Powers and roots could and, of course, still can be found by logs:

3 *To find the power of a number, multiply the log by the power*

if $N = 10^x$, then $N^n = 10^{nx}$
$\log N^n = nx$

4 *To find the root of a number, divide the log by the root index*

if $N = 10^x$, then $\sqrt[n]{N} = \sqrt[n]{10^x} = 10^{x/n}$
$\log \sqrt[n]{N} = x/n$

Test and problems 2

Multiple choice test: MT 2

Answer block:

Question No.	0	1	2	3	4	5	6	7	8
Answer	b								

Enter your answer, that is a, b, c or d in the column under the question number in the answer block above.
Note that question Qu. 0 has already been worked out and the answer inserted.

Qu. 0 The logarithm of a number may be defined as the inverse of $a^x = y$ as $\log_a y = x$. Using this definition find the logarithm of 1 000 000 to base 10.
Ans (a) 10 (b) 6 (c) 1000 (d) 2

Solution

1 000 000 expressed in indicial form to base 10 is

one million $= 10 \times 10 \times 10 \times 10 \times 10 \times 10$
$= 10^6$
hence $\log_{10} 1\,000\,000 = 6$
So the correct answer is (b) and we insert 'b' under Qu. No. 0 in the answer block

---

Now carry on with the test.

Qu. 1 Evaluate $36^{\frac{1}{2}}$
Ans (a) 6 (b) -6 (c) ± 6 (d) 3

Qu. 2 Evaluate $125^{\frac{1}{3}}$
Ans (a) 5 (b) 25 (c) -5 (d) $\frac{1}{5}$

Qu. 3 Which of the following has the greatest value:
Ans (a) 25^0 (b) $5^{\frac{1}{4}}$ (c) 0.5^{-1} (d) 5^{-2}

Qu. 4 Evaluate $8^{\frac{1}{2}} \times 8^{\frac{1}{2}} \times 64^{\frac{1}{3}}$
Ans (a) 8 (b) 16 (c) 32 (d) 48

Qu. 5 Evaluate $10^4 \div 10^6$
Ans (a) 100 (b) 10^{10} (c) $\frac{1}{10}$
(d) 10^{-2}

Qu. 6 Determine $\log_2 128$
Ans (a) 7 (b) 64 (c) 16 (d) 28

Qu. 7 If $\log_{10} 2 = 0.3010$ and $\log_{10} 3 = 0.4771$, find $\log_{10} 6$
Ans (a) 0.1761 (c) 0.6020
(b) 6 (d) 0.7781

Qu. 8 Decibels are units employed in science and engineering to measure sound intensity. The ratio of two sound levels P_1 and P_2 expressed in decibels is 60 dB:

$$10 \log_{10}(P_1/P_2) = 60\,\text{dB}$$

If $P_1 = 1$ unit, determine P_2
Ans (a) 60 (b) 6 (c) 100
(d) 10^{-6}

Problems 2

1 Evaluate the following:
(a) $81^{\frac{1}{4}}$ (b) $144^{\frac{1}{2}}$ (c) $216^{\frac{1}{3}}$
(d) $25^{\frac{3}{2}}$ (e) $125^{\frac{2}{3}}$ (f) $100^{\frac{5}{2}}$

2 Determine the values of
(a) 4^{-1} (b) 4^{-2} (c) 4^{-3}
(d) $25^{-\frac{1}{2}}$ (e) $49^{-\frac{1}{2}}$ (f) $64^{-\frac{3}{2}}$

3 Express the following in simplest form:
(a) $(10^2)^3$ (b) $(10^3)^{-2}$ (c) $(100)^{-\frac{3}{2}}$

4 Evaluate
(a) $3^{\frac{1}{4}} \times 3^{\frac{3}{4}}$ (b) $5^{\frac{1}{2}} \div 5^{-\frac{3}{2}}$ (c) $10^2 \times 10^3 \times 10^{-5}$

5 Evaluate
(a) $(6.2 \times 10^{-1.5}) \div (3.1 \times 10^{2.5})$
(b) $3.69 \times 10^{-6} \times 10^5$
(c) $7^{-\frac{1}{2}} \div 7^{\frac{1}{2}}$
(d) $9^{-\frac{1}{5}} \times 9^{\frac{4}{5}} \div 9^{\frac{3}{5}}$

6 Determine the following logarithms
(a) $\log_2 16$ (b) $\log_3 81$ (c) $\log_{10} 1000$
(d) $\log_5 25$ (e) $\log_{10} 0.01$ (f) $\log_8 8$

7 Determine y for the following
(a) $\log_2 y = 6$ (b) $\log_3 y = -2$
(c) $\log_{10} y = 6$

8 If $\log_{10} 6 = 0.7781$ and $\log_{10} 2 = 0.3010$, determine
(a) $\log_{10} 12$ (b) $\log_{10} 3$ (c) $\log_{10} 4$

3 Checking calculations and making approximations

General learning objectives: to ensure answers to numerical problems are reasonable.

3.1 Expressing a number correct to a given number of decimal places

Since we often meet decimal numbers containing far more digits than is necessary, it is common practice to state numbers 'correct to a given number of decimal places'. The number of places in a decimal is the total number of digits including any zeros after the decimal point. For example:

46.5 has one decimal place (only one digit after the decimal point)
179.26 has two decimal places
0.438 has three decimal places
0.001 also has three decimal places
3636.3334 has four decimal places

In order to express a number correct to a specified number of decimal places we use the following rule:

> The last 'specified' place is unchanged if the digit immediately following it is 4 or less; if this is 5 or more then the last 'specified' digit is increased by 1.

For example:

24.5739 specified to one decimal place is 24.6 since the digit immediately following the first decimal place, the 5, is a 7. Hence the 5 is increased by 1 to 6.

24.5739 specified to two decimal places is 24.57; the second digit, 7, is unchanged since it is followed by 3.

24.5739 specified to three decimal places is 24.574 the 3 is increased by 1 to 4 because it is followed by 9.

Examples

1 Specify the following numbers correct to one decimal place:

	22.23	672.99	0.067	10.049	7.008
Ans	22.2	673.0	0.1	10.0	7.0

2 The following results were obtained using an electronic calculator

$x = 4.7081491 \qquad y = 785.96388$
$z = 0.00098778$

The results are required accurate to two decimal places. Specify x, y and z to two decimal places

Ans $x = 4.71 \qquad y = 785.96 \qquad z = 0.00$

3 *Terminating, non-terminating and recurring decimals*
Many numerical quantities cannot be expressed exactly in decimal form. The constant π which is the ratio of the perimeter (the circumference) of a circle is an example. To nine decimal places, $\pi = 3.141592654$ and even though this is more than accurate for most applications, mathematicians have worked out π to several hundred decimal places.

Many fractions can be expressed exactly in decimal form, e.g.

$$\frac{1}{2} = 0.5, \quad \frac{1}{4} = 0.25, \quad \frac{1}{20} = 0.05$$

and in these cases we have a terminating decimal. Many fractions, however, cannot be expressed precisely in decimal form, e.g.

$$\frac{1}{3} = 0.3333;\ldots \qquad \frac{1}{7} = 0.142857142\ldots$$

and in these cases we have non-terminating decimals. Where non-terminating decimals have a repeating pattern of digits they are known as recurring decimals. Recurring decimals may be written with a dot or dots over the digits that repeat, e.g.

$\frac{1}{3} = 0.3333\ldots$ may be written as $\frac{1}{3} = 0.\dot{3}$

$\frac{1}{11} = 0.090909\ldots$ may be written as

$\frac{1}{11} = 0.\dot{0}\dot{9}$

Problems: (a) express π accurate to 4 decimal places;

(b) convert the fractions $\frac{3}{8}$ and $\frac{2}{9}$ into decimals, specifying the result accurate to 3 decimal places.

Solutions

(a) $\pi = 3.1416$ (to 4 decimal places)

(b) Remember: to convert a fraction into a decimal we divide the numerator (top number) by the denominator (bottom number) of the fraction, so

```
3      .375
- =  8)3.000
8      24
        60
        56
         40
         40
         ..
```

So $\frac{3}{8} = 0.375$ is a terminating decimal and can be expressed exactly to 3 decimal places.

```
2      .222...
- =  9)2.0000
9      18
        20
        18
         20
         18
          20
```

So clearly $\frac{2}{9}$ is a recurring decimal when converted,

$\frac{2}{9} = 0.\dot{2} = 0.222$ (to 3 decimal places).

3.2 Expressing a number correct to a given number of significant figures

The number of significant figures used to express a numerical quantity is the total number of digits obtained by counting from the left starting at the first non-zero digit, that in the case of decimal fractions we ignore any leading zeros following the decimal point.

For example,

0.057 has 2 significant figures (ignore the leading 0.0; 5 is the 1st and 7 the 2nd significant figure)

5683.2 has 5 significant figures
13241 has 5 significant figures
0.1 has 1 significant figure
0.006 has 1 significant figure

To express a number correct to a specified number of significant figures we employ the following rule:

The last required significant figure is unchanged if the digit following it is 4 or less;
the last required significant figure is increased by 1 if the digit following it is 5 or more;
all digits following the last significant figure (if any) become zeros.

For example,

33.34 expressed to 1 significant figure is 30
expressed to 2 significant figures is 33
expressed to 3 significant figures is 33.3

146 789 expressed to 1 significant figure is 100 000
expressed to 2 significant figures is 150 000
expressed to 3 significant figures is 147 000
expressed to 4 significant figures is 146 800

0.043 28 expressed to 1 significant figure is 0.04
expressed to 2 significant figures is 0.043
expressed to 3 significant figures is 0.0433

Examples

1 Specify the following correct to one significant figure

	19	784	449	0.62	0.047	0.000 423
Ans	20	800	400	0.6	0.05	0.0004

2 Specify the following correct to three significant figures.

	162.4	79.05	876 439	0.007 296 1
Ans	162	79.1	876 000	0.00730

3.3 Checking the validity and feasibility of solutions: approximations

Whatever method we employ to find solutions to a problem it is essential to apply our commonsense and practical experience to ensure the solutions are at the very least feasible and valid ones. The following points are useful to bear in mind:

1. Make a rough check on calculations. Use approximations.
2. Limit the results to a realistic degree of accuracy.
3. Alway check the validity of a result. Is it feasible?
4. Check that you have the correct data to solve the problem and, especially in scientific and engineering problems, that all the quantities are in the correct units.

The following examples illustrate the ideas of making approximations to check results and also to test the validity of solutions.

Examples

1 The calculation: $(52.67 \times 4.6) - (49.2 \times 2.9)$ is worked out on a calculator by a student who gives the answer as 99.602. Is this answer approximately correct?

Check

A rough check can be made by writing each term in the calculation accurate to one significant figure:

$$52.67 \times 4.6 \approx 50 \times 5 = 250$$
$$49.2 \times 2.9 \approx 50 \times 3 = 150$$

so on subtracting, $250 - 150 = 100$

Hence the answer is approximately correct. [Note the symbol $\approx$ means 'approximately equal to']

2 Evaluate $\dfrac{(8.04 \times 3.01) - (4.97 \times 2.95)}{50.6 \times 2.2 \times 6.8}$ by approximating each term to an accuracy of one significant figure, and quoting the result to one significant figure.

Solution

First work out the numerator, approximating each term to one significant figure, i.e.

$$(8 \times 3) - (5 \times 3) = 24 - 15 = 9$$

Likewise, the denominator:

$$50 \times 2 \times 7 = 700$$

and on dividing the approximate numerator by the approximate denominator,

$$\frac{9}{700} = 700\overline{)9.00} \quad .012\ldots$$

```
       .012...
  700)9.00
      700
      ---
      200
      140
      ---
       60
```

so the approximate result to one significant figure is 0.01 *Ans*

The result worked out by calculator is 0.0126013, so in this case our approximate result of $\frac{9}{700} \approx 0.012$ is not too far out. Note, however, approximate calculations do not necessarily give results accurate to 1 significant figure although they will provide at the very worst an estimate of the order of magnitude of the result.

3 The annual interest on a loan of £1400 where the interest is charged at 15% is quoted as £2100. Is this quotation correct?

Check

Obviously not. Commonsense tells us that an annual interest of £2100 on a loan of £1400 is clearly greater than the loan itself and therefore incorrect.

$$10\% \text{ interest on } £1400 = £1400 \times \frac{10}{100}$$
$$= £140$$

$$5\% \text{ interest on } £1400 = \frac{1}{2} \times £140 = £70$$

so the interest charged at 15% should be £140 + £70 = £210

4 You arrive at an overseas airport and exchange £50 to franc-marks (fm). The official exchange rate is £1 = fm 587.2 you receive fm 20,000. Is this transaction approximately correct?

Solution

Since $£1 = \text{fm } 587.2$

$$£50 = \text{fm } 587.2 \times 50$$
$$\approx \text{fm } 600 \times 50 = \text{fm } 30{,}000$$

so even allowing for the agent's commission that should not amount to more than a few per cent and the fact that we have approximated 587.2 to 600, an error of the order of fm 10,000 has been made in the exchange office's favour!

5 It is often necessary to make a quick check of a list of items such as a bill, weights, etc. Check that the following is approximately correct by rounding items to the nearest Pound.

Actual bill list.	*Solution: list rounded to nearest £*
£7.72	£8
£5.40	£5
£0.35	0
£162.39	£162
£48.44	£48
£5.99	£6
£11.75	£12
£432.98	£433
£675.02	£674

The summation of our approximate bill is only about £1 out so the actual bill is approximately correct.

6 In the advertising literature of a family saloon car the petrol tank capacity is quoted as 180 litres, the petrol consumption as 10 km/litre (28 m p g) and the typical full tank range of 4000 km.

[4.56 litres = 1 gallon, $1 \text{ km} = \frac{5}{8}$ mile; km is the abbreviation for kilometres and m p g for miles per gallon]

Are these facts feasible? If not which one(s) are very suspect?

Solution

Petrol tank capacity: $180 \text{ litres} \approx 180/4.6 \approx 40$ gallons. Such a large capacity is highly unlikely for a normal car. Petrol in the UK costs about £0.4 per litre or £1.80 per gallon so the cost to fill-up would be over £70 (£180 × 0.4)! Hence this additional practical or rather economic point would indicate a printing error in the literature.

The petrol consumption at 10 km/litre (28 m p g) seems reasonable. You would expect to cover 10 km (about 6 miles) on a litre of petrol and this is the same as 28 miles per gallon.

The range, however, is clearly not plausible under any circumstances: $4000 \text{ km} = 4000 \times \frac{5}{8} = 2500$ miles would more than take you from the UK to Spain and back on one tank of petrol and even if your tank capacity were 180 litres your range would be only $180 \times 10 = 1800$ km or $40 \times 28 \approx 1200$ miles.

Test and problems 3

Multiple choice test: 3

Answer block:

Question No.	0	1	2	3	4	5	6	7	8
Answer	d								

Enter your answer, that is a, b, c or d in the column under the question number in the answer block above. Note that question Qu. 0 has already been worked out and the answer inserted.

Qu. 0 Which of the following answers best approximates the value of the expression

$$\frac{38.93 \times 42.84}{6.02}$$ to two significant figures.

Ans (a) 300 (b) 30 (c) 28 (d) 280

Solution

First using approximate cancelling and then rounding, we have

$$\frac{38.93 \times 42.84}{6.02} \approx \frac{38.93 \times \overset{7}{\cancel{42.84}}}{\underset{1}{\cancel{6.02}}}$$

$$\approx 40 \times 7 = 280$$

so the correct answer is (d) and we insert 'd' under Qu. No. 0 in the answer block

---

Now carry on with the test.

Qu. 1 Specify 49.9987 to two decimal places.

Ans (a) 50.00 (b) 49
(c) 49.90 (d) 49.99

Qu. 2 Specify the number 0.054752 to 3 significant figures.

Ans (a) 5.475×10^{-2} (b) 0.055
(c) 0.0548 (d) 0.0547

Qu. 3 Evaluate the expression $9.76 \times 98.6 \times 999.6$ to an accuracy of one significant figure.

Ans (a) 100 000 (b) 9.5×10^4
(c) 1 000 000 (d) 900 000

Qu. 4 The distance of the sun from the earth is approximately 93 000 000 miles. Light travels at the speed of 186 000 miles per second. Which of the following answers is feasible for the time taken for light leaving the sun to reach the earth.

Ans (a) 1 hour (b) 500 seconds
(c) 10 minutes (d) 1 day

Qu. 5 A steel bar has a volume of 2×10^{-3} cubic metres. One cubic metre of steel is known to weigh 7700 kilograms. Determine the weight of the bar in kilograms.

Ans (a) 15.4 (b) 154
(c) 77 (d) 15.4×10^3

Qu. 6 An estimate for the number of millimetres in 1 mile, given the data: 1 mile = 63 360 inches and 1 inch = 25.4 mm, is made and the estimates are given below. Which is the most reasonable?

Ans (a) 1 000 000 (b) 1 600 000
(c) 2500 (d) 10^7

Qu. 7 A computer is capable of executing 1 million operations per second. How long does it take to complete a task involving 2000 operations?

Ans (a) 2 seconds (b) 500 seconds
(c) 0.002 seconds (d) 2 minutes

Qu. 8 The number of seconds in a year is approximately

Ans (a) 5×10^6 (b) 20×10^6
(c) 32×10^6 (d) 530 000

Problems 3

1 Specify the following to two decimal places:
(a) 36.785 (b) 0.0739
(c) 487.999 (d) 0.000 432

2 Specify the following correct to four significant figures:
(a) 64.537 (b) 579666
(c) 0.000 4779 (d) 2.718 28

3 Determine approximate values for the following calculations by rounding numbers (to 1 or 2 significant figures) and rough cancelling:

(a) $\dfrac{29.75 \times 6.28}{5.867}$ (b) $\dfrac{98.67 \times 121.79}{37.63 \times 19.81}$

(c) $\dfrac{(17.25 - 6.91)}{(5.05 - 3.06)}$

(d) $\dfrac{(7.77 + 3.27)(4.76 - 2.81)}{(28.62 - 18.0)(5.78 - 2.31)}$

After reading Chapter 4 (use of Tables) or Chapter 5 (use of calculator) calculate the exact values.

4 Determine approximate values for the following calculations, first converting individual terms to standard form:

(a) $\dfrac{5762 + 6344}{799 + 311}$ (b) $\dfrac{678 \times 402 \times 598}{0.76 \times 0.52}$

(c) $\dfrac{2000(4500 + 6200)}{500 \times 1200}$

(d) $\sqrt{\left(\dfrac{692\,000 + 110\,000}{145 \times 1390}\right)}$

5 The following table gives the relationship between some commonly used Imperial or British units and the corresponding metric units (International SI units):

British units	Metric
1 lb	= 0.4536 kg (kilograms)
1 inch	= 25.4 mm (millimetres)
1 gallon	= 4.546 litres
1 horsepower	= 746 W (watts)

Use this table to estimate approximately:
(a) 12 inches in terms of millimetres
(b) 1 mile (= 1760 yards) in terms of metres
(c) 1 ton (= 2240 lb) in terms of kilograms
(d) 1 ounce (16 oz = 1 lb) in terms of grams
(e) 1 pint (8 pints = 1 gall) in terms of litres
(f) 10 kilowatts in terms of horsepower
[Remember 1 m = 1000 mm, 1 kg = 1000 g, 1 kW = 1000 W]

6 The following calculations have been worked out to an accuracy of one significant figure. State whether the results are feasible or not, and if not give your approximation.
(a) $(4.6 \times 10^{-3}) - (2.9 \times 10^{-4}) \approx 2 \times 10^{-7}$
(b) $(5.5 \times 10^{6}) \div (2.7 \times 10^{3}) \approx 2 \times 10^{3}$
(c) $\dfrac{7.2 - 2.9}{0.021} \approx 2000$
(d) The annual interest on £10,000 invested at 7.25% p.a. is £700
(e) The reciprocal of $86 \approx 0.01$
(f) $(8.6^2 + 27)^2 \approx 1000$

4 Using mathematical tables and charts

General learning objectives: to understand and to use mathematical tables and charts.

4.1 Introduction: tables of mathematical function values

Booklets of mathematical tables are readily available, which contain tabulated values of the more commonly used mathematical functions such as logarithms (base 10) and anti-logarithms (10^x), squares, square roots, reciprocals and trigonometric functions (sines, cosines and tangents).

The values of the functions listed in the tables are usually given to four significant figures and in such cases the tables are known as 4-figure tables. Four-figure accuracy is usually more than sufficient for most calculations and limiting to four figures enables the values to be tabulated in rows and columns. For greater accuracy 7-figure tables are available but are far more cumbersome to use. It is, however, interesting to note that a good-quality electronic calculator supplies at the press of a key all the above functions and usually several extra ones to an accuracy of least eight or more significant figures.

4.2 Using tables to find squares, square roots and reciprocals

In this section we explain how 4-figure mathematical tables are used to determine the squares, square roots and reciprocals of numbers.

4.2.1 Finding squares

Four-figure 'squares' tables list the squares of numbers giving the square to an accuracy of four significant figures. No information can be gained directly from the tables as to the order of magnitude of the square since the tables do not include the decimal point. It is up to the user to put it in. This is usually the case with most tables. Hence rough order of magnitude calculations should always be made. The procedure is explained in the following examples.

Examples

1 Determine the squares of the following numbers:
(a) 1.546 (b) 1546 (c) 0.01546

Procedure

A portion of the squares tables, relevant to our present problem is reproduced on next page 21.

Locate 1546 on the table – at this stage do not worry about any decimal points or indicial form multipliers. This is done by placing a ruler along the 15 row and moving to the third significant digit column – the 4 column.

Read off the square of 154, i.e. 2372 which is circled in the table.

Next move to the fourth significant figure column – the 6 column, in our case and read off the 19. This value is known as the mean difference and is to be added to 2372 to find the square of 1546, i.e.

$$154^2 \rightarrow 2372$$
$$\text{mean difference } 6 \rightarrow \quad 19$$
$$\text{add, } 1546^2 \rightarrow 2391$$

So with, as yet no information about the position of the decimal point, we have from the tables

$$1546^2 = 2391 \text{ to four significant figures}$$

Consider now positioning the decimal point to evaluate,

(a) 1.546^2

Rough calculation: 1.546^2 must lie between 1^2 and 2^2, i.e. between 1 and 4, or even better
$$1.546^2 \approx 1.5 \times 1.5 = 2.25$$

SQUARES

	0	1	2	3	4	5	6	7	8	9	1 2 3	4 5 6	7 8 9
10	1000	1020	1040	1061	1082	1103	1124	1145	1166	1188	2 4 6	8 10 13	15 17 19
11	1210	1232	1254	1277	1300	1323	1346	1369	1392	1416	2 5 7	9 11 14	16 18 21
12	1440	1464	1488	1513	1538	1563	1588	1613	1638	1664	2 5 7	10 12 15	17 20 22
13	1690	1716	1742	1769	1796	1823	1850	1877	1904	1932	3 5 8	11 13 16	19 22 24
14	1960	1988	2016	2045	2074	2103	2132	2161	2190	2220	3 6 9	12 14 17	20 23 26
15	2250	2280	2310	2341	2372	2403	2434	2465	2496	2528	3 6 9	12 15 19	22 25 28
16	2560	2592	2624	2657	2690	2723	2756	2789	2822	2856	3 7 10	13 16 20	23 26 30
17	2890	2924	2958	2993	3028	3063	3098	3133	3168	3204	3 7 10	14 17 21	24 28 31
18	3240	3276	3312	3349	3386	3423	3460	3497	3534	3572	4 7 11	15 18 22	26 30 33
19	3610	3648	3686	3725	3764	3803	3842	3881	3920	3960	4 8 12	16 19 23	27 31 35

First two non-zero digits of number to be squared

Third digit columns

Fourth digit (mean difference) columns

so we see we have a single digit in front of the decimal point:

$$1.546^2 = 2.391 \text{ (to 4 significant figures)}$$

Ans

(b) 1546^2

Converting to standard form,

$1546 = 1.546 \times 10^3$

so $1546^2 = (1.546 \times 10^3)^2$

$= 1.546^2 \times 10^6$

$= 2.391 \times 10^6$ or 2 391 000 *Ans*

(c) 0.01546^2

$0.01546 = 1.546 \times 10^{-2}$

so $0.01546^2 = (1.546 \times 10^{-2})^2$

$= 1.546^2 \times 10^{-4}$

$= 2.391 \times 10^{-4}$ or 0.000 2391

Ans

2 Determine, using 4-figure square tables:
(a) 962^2 (b) 0.4624^2

Answers

(a) From the square tables,

$962^2 = 9254$ with position of decimal point to be determined.

Rough calculation:

$962^2 \approx 1000 \times 1000 = 1\,000\,000$

i.e. 962^2 should be just below a million, hence

$962^2 = 925\,400$ (to 4 significant figures)

(b) From the tables

$4624^2 = 2134 + 4 = 2138$ with the position of the decimal point for the case of 0.4624^2 to be determined. Rough calculation:

$0.4624^2 \approx 0.5 \times 0.5 = 0.25$,

hence

$0.4624^2 = 0.2138$

4.2.2 Finding square roots

When you look up square root tables you will find two values listed. It is our task to select the relevant value and then as always with tables, to insert the decimal point in the correct position. This is accomplished by finding an approximate value and matching this to one of the pair of values given in the table. The procedure is illustrated in the following examples.

Examples

1 Find the square root of (a) 35.36 (b) 3.536
(c) 0.003 536 (d) 35 360

	0	1	2	3	4	5	6	7	8	9	1 2 3	4 5 6	7 8 9
33	1817	1819	1822	1825	1828	1830	1833	1836	1838	1841	0 1 1	1 1 2	2 2 2
	5745	5753	5762	5771	5779	5788	5797	5805	5814	5822	1 2 3	3 4 5	6 7 8
34	1844	1847	1849	1852	1855	1857	1860	1863	1865	1868	0 1 1	1 1 2	2 2 2
	5831	5840	5848	5857	5865	5874	5882	5891	5899	5908	1 2 3	3 4 5	6 7 8
35	1871	1873	1876	1879	1881	1884	1887	1889	1892	1895	0 1 1	1 1 2	2 2 2
	5916	5925	5933	5941	5950	5958	5967	5975	5983	5992	1 2 2	3 4 5	6 7 8
36	1897	1900	1903	1905	1908	1910	1913	1916	1918	1921	0 1 1	1 1 2	2 2 2
	6000	6008	6017	6025	6033	6042	6050	6058	6066	6075	1 2 2	3 4 5	6 7 7
37	1924	1926	1929	1931	1934	1936	1939	1942	1944	1947	0 1 1	1 1 2	2 2 2
	6083	6091	6099	6107	6116	6124	6132	6140	6148	6156	1 2 2	3 4 5	6 7 7
38	1949	1952	1954	1957	1960	1962	1965	1967	1970	1972	0 1 1	1 1 2	2 2 2
	6164	6173	6181	6189	6197	6205	6213	6221	6229	6237	1 2 2	3 4 5	6 6 7
39	1975	1977	1980	1982	1985	1987	1990	1992	1995	1997	0 1 1	1 1 2	2 2 2
	6245	6253	6261	6269	6277	6285	6293	6301	6309	6317	1 2 2	3 4 5	6 6 7

Procedure

The relevant portion of the square root tables is shown above.

The values corresponding to the square roots for our case are found by locating the 35 row and 3 and 6 columns. These values are 'ringed' in the tables above, and are given by

$$\sqrt{3536} \rightarrow 1879 + 2 = 1881$$
$$5941 + 5 = 5946$$

(a) $\sqrt{35.36}$

Since $35.36 \approx 36$, $\sqrt{35.36} \approx 6$
so the lower value in the table is the appropriate one,

$$\sqrt{35.36} = 5.946 \quad \textit{Ans}$$

(b) $\sqrt{3.536}$

Since $3.536 \approx 4$, $\sqrt{3.536} \approx \sqrt{4} = 2$ and so the top value is selected, i.e. $\sqrt{3.536} = 1.881$ *Ans*

(c) $\sqrt{0.003\,536}$

A useful rule for finding an approximate square root so that we know where to place the decimal point is firstly to pair off digits from the decimal point, then to find the nearest whole number square root of the most significant non-zero digit pair, and finally then add zeros, one each for all of the other pairs, e.g.

$$\sqrt{0.|00|\ 35|36}$$
↑
0. 0 6 so $\sqrt{0.003\,536} \approx 0.06$

and hence from lower value in table,

$$\sqrt{0.003\,536} = 0.05946 \text{ (to 4 significant figures)} \quad \textit{Ans}$$

(d) $\sqrt{35\,360}$

$$\sqrt{3|53|60|}.$$
↑
1 0 0 so $\sqrt{35\,360} \approx 100$

and using upper value in table,

$$\sqrt{35\,360} = 188.1$$

2 Determine the square roots of (a) 399 800 (b) 0.3312

(a) $\sqrt{39|98|00|}$
↑
6 0 0 so approximate square root is 600.

RECIPROCALS SUBTRACT

	0	1	2	3	4	5	6	7	8	9	1 2 3	4 5 6	7 8 9
5.5	0.1818	0.1815	0.1812	0.1808	0.1805	0.1802	0.1799	0.1795	0.1792	0.1789	0 1 1	1 2 2	2 3 3
5.6	0.1786	0.1783	0.1779	0.1776	0.1773	0.1770	0.1767	0.1764	0.1761	0.1757	0 1 1	1 2 2	2 3 3
5.7	0.1754	0.1751	0.1748	0.1745	0.1742	0.1739	0.1736	0.1733	0.1730	0.1727	0 1 1	1 2 2	2 2 3
5.8	0.1724	0.1721	0.1718	0.1715	0.1712	0.1709	0.1706	0.1704	0.1701	0.1698	0 1 1	1 1 2	2 2 3
5.9	0.1695	0.1692	0.1689	0.1686	0.1684	0.1681	0.1678	0.1675	0.1672	0.1669	0 1 1	1 1 2	2 2 3
6.0	0.1667	0.1664	0.1661	0.1658	0.1656	0.1653	0.1650	0.1647	0.1645	0.1642	0 1 1	1 1 2	2 2 3
6.1	0.1639	0.1637	0.1634	0.1631	0.1629	0.1626	0.1623	0.1621	0.1618	0.1616	0 1 1	1 1 2	2 2 2
6.2	0.1613	0.1610	0.1608	0.1605	0.1603	0.1600	0.1597	0.1595	0.1592	0.1590	0 1 1	1 1 2	2 2 2
6.3	0.1587	0.1585	0.1582	0.1580	0.1577	0.1575	0.1572	0.1570	0.1567	0.1565	0 0 1	1 1 1	2 2 2

Note that the mean difference values in this column must be SUBTRACTED

Hence square root corresponds to the lower value in the tables, i.e. $6317 + 6 = 6323$ and inserting the decimal point,

$$\sqrt{399\,800} = 632.3 \text{ (to 4 significant figures)}$$

Ans

(b) $\sqrt{0.|33|12}$

$\downarrow$

0. 5 0, $\sqrt{0.3312} \approx 0.5$

and taking lower alternative from tables, $5753 + 2 = 5755$
so $\sqrt{0.3312} = 0.5755$ (to 4 significant figures)

Ans

4.2.3 Finding reciprocals

Finding reciprocals using four-figure tables is similar to that of finding squares or square roots except the mean difference, for the fourth digit, in the far right column must be subtracted.

Example

Find the reciprocals of
(a) 5.843 (b) 55.67 (c) 0.006 309

Procedure

The relevant portion of the reciprocal tables covering these values is shown above.

(a) Reciprocal of 5.843
5.84 is located along the 5.8 row, central 4 column, and is shown ringed in the above table

$$5.84^{-1} = 0.1712$$

The fourth significant digit 3, located in the far right column has a mean difference value of 1, in fact 0.0001, which must be subtracted, so

$$5.843^{-1} = \begin{array}{r} 0.1712 \\ -\quad 1 \\ \hline 0.1711 \\ \hline \end{array}$$

i.e. $5.843^{-1} = 0.1711$ *Ans*

(b) 55.67^{-1}

Rough calculation $55.67^{-1} = \dfrac{1}{50} = 0.02$

Now look up the reciprocal of 5567 without worrying about the decimal point,

$$5567^{-1} = \begin{array}{r} 0.1799 \\ -\quad 2 \\ \hline 0.1797 \\ \hline \end{array}$$

and finally inserting the decimal point, being guided as to its position by our rough calculation, we have

$$55.67^{-1} = 0.01797 \quad \textit{Ans}$$

(c) $0.006\,309^{-1}$

Rough calculation

$$0.006\,309^{-1} = 0.006^{-1}$$

$$= \frac{1}{0.006} = \frac{1000}{6} \approx 170$$

From tables $6.309^{-1} =$

$$\begin{array}{r} 0.1587 \\ -\quad 2 \\ \hline 0.1585 \\ \hline \end{array}$$

and using the rough calculation to determine the decimal point position, we obtain

$$0.006\,309^{-1} = 158.5 \quad \textit{Ans}$$

4.3 Using log tables to aid calculations

4.3.1 Introduction

We can use tables of common logarithms (logs to base 10) to carry out the multiplication and division of numbers and also to evaluate powers, roots and reciprocals.

Before the widespread availability of reasonably priced electronic calculators, log tables were used extensively to carry out calculations. Compared to a calculator their use is far more cumbersome, much slower and for the 4-significant tables much less accurate. However, they can be useful in emergencies – such as when your calculator breaks down or its batteries are flat!

The most common form of logarithms and the ones most commonly used in numerical calculation work are logarithms to base 10. As explained in section 2.4, the logarithm of a number N to base 10 is the power to which the base 10 must be raised to equal N, i.e.

if $\quad N = 10^x$

then the index x is the logarithm to base 10 of N,

$$x = \log_{10} N$$

The actual value of the logarithm, that is x, is found by consulting log tables. The 10 is usually omitted when we are working with common logarithms. For example, the common logarithm of 7 to base 10 is written as,

$$\log 7 = 0.8451$$

and using the definition for a logarithm, we can state

$$7 = 10^{0.8451}$$

$$\begin{array}{ccc} \uparrow & \uparrow & \uparrow \\ \text{number} & \text{base} & \text{log} \end{array}$$

Using logs simplifies calculation work as the log process essentially reduces

multiplication into an addition process
division into a subtraction process
determining powers into a multiplication process
determining roots into a division process

Since we can apply the rules of indices (see sections 2.2 and 2.4):

$10^{x_1} \times 10^{x_2} = 10^{x_1+x_2}$ to multiply, add logs
$10^{x_1} \div 10^{x_2} = 10^{x_1-x_2}$ to divide, subtract logs
$(10^x)^n = 10^{x\times n}$ multiply log by power n
$\sqrt[n]{(10^x)} = 10^{x/n}$ divide log by n to find root

We use log tables to find the logarithms, x_1 and x_2, of the numbers. We use antilog tables to reverse the process, that is to find the number corresponding to the log result. Antilog tables list values of 10^x for values of x. The process of making a calculation using logs is summarized in Figure 4.1.

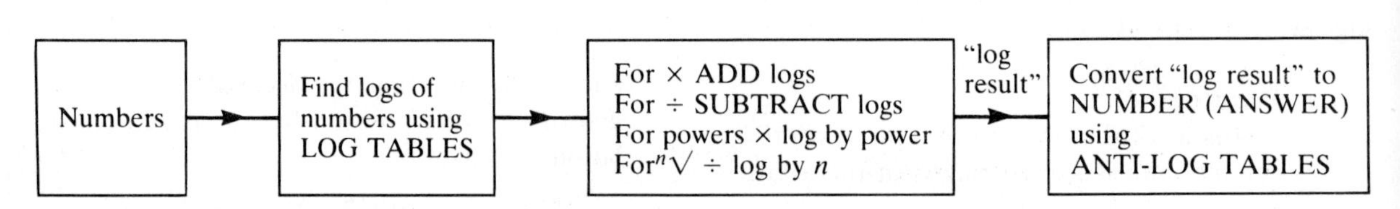

Figure 4.1 Block diagram showing the basic steps involved in using log tables to effect calculations.

	0	1	2	3	4	5	6	7	8	9	1 2 3	4 5 6	7 8 9
10	0.0000	0.0043	0.0086	0.0128	0.0170	0.0212	0.0253	0.0294	0.0334	0.0374	4 8 12	17 21 25	29 33 37
11	0.0414	0.0453	0.0492	0.0531	0.0569	0.0607	0.0645	0.0682	0.0719	0.0755	4 8 11	15 19 23	26 30 34
12	0.0792	0.0828	0.0864	0.0899	0.0934	0.0969	0.1004	0.1038	0.1072	0.1106	3 7 10	14 17 21	24 28 31
13	0.1139	0.1173	0.1206	0.1239	0.1271	0.1303	0.1335	0.1367	0.1399	0.1430	3 6 10	13 16 19	23 26 29
14	0.1461	0.1492	0.1523	0.1553	0.1584	0.1614	0.1644	0.1673	0.1703	0.1732	3 6 9	12 15 18	21 24 27
15	0.1761	0.1790	0.1818	0.1847	0.1875	0.1903	0.1931	0.1959	0.1987	0.2014	3 6 8	11 14 17	20 22 25
16	0.2041	0.2068	0.2095	0.2122	0.2148	0.2175	0.2201	0.2227	0.2253	0.2279	3 5 8	11 13 16	18 21 24
17	0.2304	0.2330	0.2355	0.2380	0.2405	0.2430	0.2455	0.2480	0.2504	0.2529	2 5 7	10 12 15	17 20 22
18	0.2553	0.2577	0.2601	0.2625	0.2648	0.2672	0.2695	0.2718	0.2742	0.2765	2 5 7	9 12 14	16 19 21
19	0.2788	0.2810	0.2833	0.2856	0.2878	0.2900	0.2923	0.2945	0.2967	0.2989	2 4 7	9 11 13	16 18 20
20	0.3010	0.3032	0.3054	0.3075	0.3096	0.3118	0.3139	0.3160	0.3181	0.3201	2 4 6	8 11 13	15 17 19
21	0.3222	0.3243	0.3263	0.3284	0.3304	0.3324	0.3345	0.3365	0.3385	0.3404	2 4 6	8 10 12	14 16 18
22	0.3424	0.3444	0.3464	0.3483	0.3502	0.3522	0.3541	0.3560	0.3579	0.3598	2 4 6	8 10 12	14 15 17
23	0.3617	0.3636	0.3655	0.3674	0.3692	0.3711	0.3729	0.3747	0.3766	0.3784	2 4 6	7 9 11	13 15 17

Figure in front of decimal point and first decimal place figure

Second decimal place columns

Mean differences columns to be added for third decimal place

4.3.2 Log and antilog tables: application to multiplication and division

Four-figure tables of common logarithms list the values of $\log_{10} N$ for N in the range 1.000 and 9.999. The first portion of a 4-figure log table is given above.

Let us use this table to find the logarithms of, for example, 1.653 and 2.078, and at the same time give some insight into the use of logs in multiplication and division.

$$\begin{array}{rl} \log 1.653 = 0.2175 & \leftarrow \text{log corresponding to 1.65} \\ 8 & \leftarrow \text{mean difference for third decimal place digit 3} \\ \hline 0.2183 & \end{array}$$

$$\begin{array}{rl} \log 2.078 = 0.3160 & \leftarrow \text{log corresponding to 2.07} \\ 17 & \leftarrow \text{mean difference for third decimal place digit 8} \\ \hline 0.3177 & \end{array}$$

i.e. $\log 1.653 = 0.2183$ so $1.653 = 10^{0.2183}$
$\log 2.078 = 0.3177$ so $2.078 = 10^{0.3177}$

So if we wished to evaluate 1.653×2.078:

$$\begin{aligned} 1.653 \times 2.078 &= 10^{0.2183} \times 10^{0.3177} \\ &= 10^{0.2183+0.3177} \end{aligned}$$

add logs $\longrightarrow$ 0.2183 + 0.3177 = 0.5360

$$= 10^{0.5360}$$

If we can now find the value of $10^{0.5360}$ we have found the resulting product essentially by an addition process.

The inverse process of finding a number knowing its logarithm is accomplished using antilog tables. These list the values of 10^x for x in the range 0.0000 to 9.999 corresponding to $N = 1$ to 9.999. A portion of antilog tables relevant to our present problem, i.e. finding $10^{0.5360}$, is given on next page 26.

The number corresponding to the antilog of 0.5360, that is, $10^{0.5360}$ is shown ringed in the antilog tables:

antilog (0.5360) = 3436
(with no decimal point information)

$10^{0.5360}$ lies between $10^0 = 1$ and $10^1 = 10$, so its value is 3.436 and hence

	0	1	2	3	4	5	6	7	8	9	1 2 3	4 5 6	7 8 9
.50	3162	3170	3177	3184	3192	3199	3206	3214	3221	3228	1 1 2	3 4 4	5 6 7
.51	3236	3243	3251	3258	3266	3273	3281	3289	3296	3304	1 2 2	3 4 5	5 6 7
.52	3311	3319	3327	3334	3342	3350	3357	3365	3373	3381	1 2 2	3 4 5	5 6 7
.53	3388	3396	3404	3412	3420	3428	3436	3443	3451	3459	1 2 2	3 4 5	6 6 7
.54	3467	3475	3483	3491	3499	3508	3516	3524	3532	3540	1 2 2	3 4 5	6 6 7
.55	3548	3556	3565	3573	3581	3589	3597	3606	3614	3622	1 2 2	3 4 5	6 7 7
.56	3631	3639	3648	3656	3664	3673	3681	3690	3698	3707	1 2 3	3 4 5	6 7 8
.57	3751	3724	3733	3741	3750	3758	3767	3776	3784	3793	1 2 3	3 4 5	6 7 8
.58	3802	3811	3819	3828	3837	3846	3855	3864	3873	3882	1 2 3	4 4 5	6 7 8
.59	3890	3899	3908	3917	3926	3936	3945	3954	3963	3972	1 2 3	4 5 5	6 7 8

$1.653 \times 2.078 = \text{antilog}\ (0.5360) = 3.436$

Consider next how log tables can be used for division, by finding, for example $1.923 \div 1.247$.

First we look up the logs of the two numbers:

log 1.923 = 0.2833 ← log corresponding to 1.92
7 ← mean difference for third decimal place digit 3
0.2840

log 1.247 = 0.0934 ← log corresponding to 1.24
24 ← mean difference for third decimal place digit 7
0.0958

i.e. $\log 1.923 = 0.2840$ so $1.923 = 10^{0.2840}$
$\log 1.247 = 0.0958$ so $1.247 = 10^{0.0958}$

Hence if we now evaluate,

$$1.923 \div 1.247 = 10^{0.2840} \div 10^{0.0958}$$
$$= 10^{0.2840 - 0.0958}$$

subtract logs → 0.2840 − 0.0958 = 0.1882

$$= 10^{0.1882}$$

and hence the answer is

$10^{0.1882} = \text{antilog}\ (0.1882)$
= 1542 ← obtained from 0.18 row, columm 8 of antilog tables
1 ← mean difference, column 2 for right of antilog tables
1.543
↑ insert decimal point here as $10^{0.1882}$ lies between 1 and 10

Thus $1.923 \div 1.247 = 1.543$

In the above two examples we have shown that:
Multiplication may be accomplished by adding logs
division may be accomplished by subtracting logs
We first look up the log values, add or subtract as appropriate, and then look up antilog tables to convert the 'log result' into the number answer. Note, however, that log tables give log values only in the range 1.000 to 9.999; antilog tables give the number result but with no information concerning the position of the decimal point.

4.3.3 Finding the log of any number

This section explains how we can find the log of any number and the antilog results, regardless of magnitude.

It is a straightforward matter to deduce the logs of numbers which are multiples of 10, simply by applying the definition of a logarithm:

If $N = 10^x$ then $\log N = x$
(N: number; x: log)

For example,

$1\,000 = 10^3$ so $\log 1\,000 = 3$
$100 = 10^2$ so $\log 100 = 2$
$10 = 10^1$ so $\log 10 = 1$
$1 = 10^0$ so $\log 1 = 0$

We can carry on the same technique for negative whole number powers of 10:

Table 4.1 Logs of multiples and submultiples of 10

Number N	Expressed as power of 10	$\log N$
10 000	10^4	4
1 000	10^3	3
100	10^2	2
10	10^1	1
1	10^0	0
0.1	10^{-1}	$\bar{1}$
0.01	10^{-2}	$\bar{2}$
0.001	10^{-3}	$\bar{3}$
0.0001	10^{-4}	$\bar{4}$
In general, when n is any integer	10^n 10^{-n}	n $\bar{n}$

$$0.1 = \frac{1}{10} = 10^{-1} \quad \text{so } \log 0.1 = -1$$

$$0.01 = \frac{1}{100} = \frac{1}{10^2} = 10^{-2} \quad \text{so } \log 0.01 = -2$$

$$0.001 = \frac{1}{1\,000} = \frac{1}{10^3} = 10^{-3} \quad \text{so } \log 0.001 = -3$$

Note that the logarithms of 0.1, 0.01, 0.001, ⋯ are negative integers. Rather than write −1, −2, −3, ⋯ we use the **bar** convention:

bar 1 written as $\bar{1}$ means −1
bar 2 written as $\bar{2}$ means −2
bar 3 written as $\bar{3}$ means −3

Table 4.1 gives a list of the logs of multiples and submultiples of 10.

Using the above information and the fact that multiplication is carried out by adding logs, we can find the log of any number. For example:

(a) $4672 = 4.672 \times 10^3$
so $\log 4672 = \underbrace{\log 4.672}_{\text{from log tables}} + \log 10^3$
$= 0.6695 + 3 = 3.6695$

(b) $0.08291 = 8.291 \times 10^{-2}$
so $\log 0.08291 = \log 8.291 + \log 10^{-2}$
$= 0.9186 + \bar{2} = \bar{2}.9186$

A general *rule* to find logs:

1 Express the number in standard form, e.g.

$7\,825\,000 = \underbrace{7.825}_{\text{mantissa}} \times 10^6$

$0.000\,6231 = \underbrace{6.231}_{\text{mantissa}} \times 10^{-4}$

2 Use log tables to find the log of the mantissa. This log will always be a decimal. 'Add' to this the 'index' of the base 10 power, e.g.

$\log 7\,825\,000 = 6.8935 \quad \log 0.000\,6231 = \bar{4}.7946$

The names **characteristic** and **mantissa** are given respectively to the whole number (integer) and the positive decimal parts of a logarithm, e.g.

$\log 567.2 = 2.7538$ (2 = characteristic, .7538 = mantissa) $\quad \log 0.1732 = \bar{1}.2385$ ($\bar{1}$ = characteristic, .2385 = mantissa)

To find the number given its logarithm, the general *rule* is

1 Use antilog tables to find the antilog of the mantissa (decimal part) of the log only.
2 Place the decimal point in the antilog number so there is a single digit in front of the decimal point.
3 The characteristic (whole number part of log) tells us the power of 10 by which the number found in 2) is to be multiplied.

For example, to find the antilog of 3.4576:

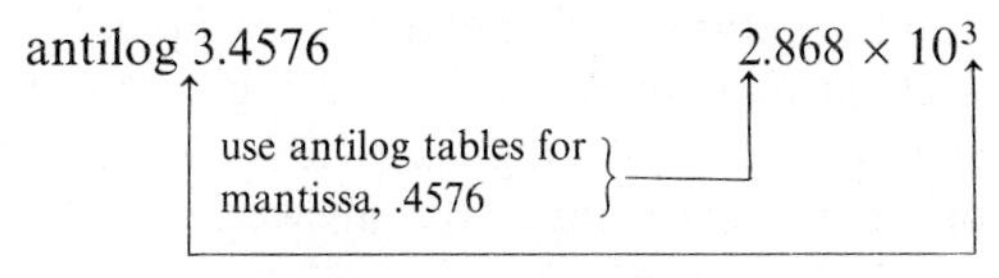

thus antilog $3.4576 = 2.868 \times 10^3$ or 2868

In simpler terms we can state the characteristic for the:

log of a number between 1 and 10 is 0.
log of a number between 10 and 100 is 1.
log of a number between 100 and 1 000 is 2.

i.e. the characteristic is *one less* than the number of digits in front of the decimal point.

log of decimal between 0.1 and 0.9999 is $\bar{1}$.
log of a decimal between 0.01 and 0.09 99 is $\bar{2}$.

Table 4.2

Number N	$\log N$
0.003	$\bar{3}.4771$
0.050	$\bar{2}.6990$
0.627	$\bar{1}.7973$
6.740	0.8287
88.880	1.9488
792.600	2.8991
1 648.000	2.2170

Table 4.3

log	antilog (10^x)
$\bar{3}.1628$	0.001 455
$\bar{2}.0030$	0.010 07
$\bar{1}.9864$	0.9692
0.6709	4.687
1.2372	17.27
2.8418	694.7
3.6246	4 213.

i.e. for decimals, the bar characteristic is *one more* than the number of zeros behind the decimal point to the first non-zero digit.

Tables 4.2 and 4.3 gives some results of finding logs and antilogs. Check that you agree.

4.4 Application of log tables: multiplication, division, powers, roots and reciprocals

The procedure for *multiplication and division* using log tables is:

1. Find the logs of the numbers using log tables, taking care to input the appropriate characteristic.
2. For *multiplication*, ADD logs.
   For *division*, SUBTRACT logs:
   log (dividend) − log (divisor).
3. Use antilog tables to convert log result to number answer, taking care to look up only the decimal part (mantissa) of the log result. The characteristic of the log result 'tells' the position of the decimal point in our answer.

Examples

1 Evaluate $526.4 \times 0.000\,824$ using 4-figure log tables.

Procedure

No.	Log.	
526.4	2.7213	add logs
0.000 824	$\bar{4}.9159$	
0.4337 ← antilog	$\bar{1}.6372$	

since characteristic = carry $1 + \bar{4} + 2 = 1 - 4 + 2 = -1 = \bar{1}$.

$526.4 \times 0.000\,824 = 0.4337$ *Ans*

It is interesting to check the accuracy of our answer, using an electronic claculator:

$$526.4 \times 0.000\,824 = 0.433\,7536 = 0.4338$$

(to 4 significant figures)

Thus the result obtained using log tables differs by 1 in the fourth significant figure. The reason for this is that 4-figure log tables only provide log values to 4 decimal places and thus, when two or more log values are added, the 'rounding' errors in each of the log values may lead to an error in the fourth decimal place in the log result and hence in the number answer. Normally we can expect an accuracy to within ± 1 in the fourth significant figure, which for most calculations is 'good enough'.

2 Evaluate $62.79 \div 279.8$.

No.	Log.	
62.79	1.7979	subtract logs
279.8	2.4469	
0.2244 ← antilog	$\bar{1}.3510$	$1 - 2 = -1 = \bar{1}$

$62.79 \div 279.8 = 0.2244$ *Ans*

[Check by calculator: 0.224 102 93 = 0.2244 (to 4 decimal places).

3 Evaluate $\dfrac{0.6821 \times 5.639 \times 792.1}{56.42 \times 0.0147}$.

Numerator

No.	Log.	
0.6821	$\bar{1}.8339$	add
5.639	0.7512	
792.1	2.8988	

Denominator

No.	Log.	
56.42	1.7515	add
0.0147	$\bar{2}.1673$	
	$\bar{1}.9188$	

No.	Log.	
	3.4839	subtract, i.e. divide num. by denom.
	$\bar{1}.9188$	
3674. ← antilog	3.5651	

So *answer* is 3674.
Check by calculator: 3 673.497 = 3 673 (to 4 s.f.).

To find POWERS, i.e. N^n

1. Look up log of number.
2. Multiply log by power, i.e. $n \times \log N$.
3. Use antilog tables to convert log result to number result.

To find ROOTS, i.e. $\sqrt[n]{N}$

1. Look up log of number.
2. Divide log by n, i.e. $(\log N) \div n$.
3. Use antilog tables to convert log result to number result.

Note: if the characteristic is a bar, always adjust this to be exactly divisible by the root n, e.g.
To divide $\bar{1}.4824$ by 2

$$\bar{1} = \bar{2} + 1 \quad \text{so} \quad \bar{1}.4824 = \bar{2} + 1.4824$$

Hence $\bar{1}.4824 \div 2 = (\bar{2} + 1.4824) \div 2$
$= \bar{1} + 0.7412 = \bar{1}.7412$

To divide $\bar{3}.6745$ by 5:

$$\bar{3}.6745 \div 5 = (\bar{5} + 2.6745) \div 5 = \bar{1}.5349$$

To find RECIPROCALS. i.e. $\frac{1}{N}$

Note that finding a reciprocal is equivalent to the division process: $1 \div N$.
So the log result is

$$\log 1 - \log N = 0 - \log N, \text{ since } \log 1 = 0$$

For example, to find the reciprocal of 21.18

No.	Log.	
1.000	0.0000	subtract
21.18	1.3259	
0.0472	$\bar{2}.6741$	← antilog

so $21.18^{-1} = 0.0472$ *Ans*

Examples

1 Evaluate using log tables

(a) 4.78^5 (b) $\sqrt[3]{57.62}$ (c) $\sqrt[4]{0.6237}$

Solutions

(a)

No.	Log.
4.78	0.6794
	× 5
2495	3.3970

(2495 ← antilog of 3.3970)

so $4.78^5 = 2495$ *Ans*

(b)

No.	Log.	
57.62	1.7606	
3.862	0.5867	← dividing 1.7606 by 3

(3.862 ← antilog of 0.5867)

so $\sqrt[3]{57.62} = 3.862$ *Ans*

(c)

No.	Log.	
0.6237	$\bar{1}.7950$	
	$= \bar{4} + 3.7950$	since $\bar{1} = \bar{4} + 3$
0.8887	$\bar{1}.9488$	← dividing $\bar{4} + 3.7950$ by 4

(0.8887 ← antilog of $\bar{1}.9488$)

so $\sqrt[4]{0.6237} = 0.8887$ *Ans*

2 Evaluate $\sqrt{(47.98)} \times 0.4634 + 2.587^{-1} \times 6.393$

Solution

It is essential to understand that we cannot use logs to add or subtract numbers. We carry out individual parts (×, ÷, powers, etc.) using logs and then effect the addition and subtraction where necessary with the number results.

No.	Log.	
47.98	1.6810	Divide by 2 to find square root
$\sqrt{(47.98)}$	0.8405	Add to multiply
0.4634	$\bar{1}.6660$	
3.210	0.5065	

(3.210 ← antilog of 0.5065)

No.	Log.	
1	0.0000	Subtract to find reciprocal
2.587	0.4128	
2.587^{-1}	$\bar{1}.5872$	Add to multiply
6.393	0.8057	
2.471	0.3929	

antilog

and finally adding the two 'number' results

$$\sqrt{(47.98)} \times 0.4634 + 2.587^{-1} \times 6.393$$
$$= 3.210 + 2.471 = 5.681 \quad Ans$$

4.5 Using trigonometric tables to find sine, cosine and tangent values

In Chapter 15 we consider the trigonometric functions: sine, cosine and tangent. These functions are extremely useful in solving problems where the calculation of lengths and angles are involved. They are also widely used in science and engineering.

Values for sine, cosine and tangent are listed in most 4-figure tables under *natural* sines, cosines and tangents for angles between 0° and 90°. Many tables also included log values so always make sure you look under the 'natural' tables if you want the actual values.

We use sine, cosine, and tangent tables in an exactly similar way but with the important difference that we are looking up values corresponding to angles expressed in degrees and minutes. Remember:

> complete circular swing corresponds to 360° one quarter of a circular swing is 90° (a right-angle) $1° = 60'$ where the ° is the symbol for degree and ′ is the symbol for minutes.

The following examples illustrate the use of 4-figure tables to determine the sine, cosine and tangent of angles.

Examples

1 Using the portion of 4-figure sine tables given below, determine the values of
(a) sin 48° (sine is normally abbreviated to sin)
(b) sin (50° 57′)
Find also the angles whose sines are (c) 0.7290 and (d) 0.8254.

Solution

(a) sin 48° = 0.7431 (located in 48° row, 0′ column)

(b) sin (50° 57′) = 0.7760 ← sin (50° 54′)

6 ← mean difference for extra 3′

0.7766 *Ans*

NATURAL SINES

	0′	6′	12′	18′	24′	30′	36′	42′	48′	54′	1′	2′	3′	4′	5′
45°	.7071	7083	7096	7108	7120	7133	7145	7157	7169	7181	2	4	6	8	10
46	.7193	7206	7218	7230	7242	7254	7266	7278	7290	7302	2	4	6	8	10
47	.7314	7325	7337	7349	7361	7373	7385	7396	7408	7420	2	4	6	8	10
48	.7431	7443	7455	7466	7478	7490	7501	7513	7524	7536	2	4	6	8	10
49	.7547	7559	7570	7581	7593	7604	7615	7627	7638	7649	2	4	6	8	9
50	.7660	7672	7683	7694	7705	7716	7727	7738	7749	7760	2	4	6	7	9
51	.7771	7782	7793	7804	7815	7826	7837	7848	7859	7869	2	4	5	7	9
52	.7880	7891	7902	7912	7923	7934	7944	7955	7965	7976	2	4	5	7	9
53	.7986	7997	8007	8018	8028	8039	8049	8059	8070	8080	2	3	5	7	9
54	.8090	8100	8111	8121	8131	8141	8151	8161	8171	8181	2	3	5	7	8
55	.8192	8202	8211	8221	8231	8241	8251	8261	8271	8281	2	3	5	7	8

Note: All sines (and cosines) between 0° and 90° have values less than one, so although the decimal point is not included except in the 0′ column it is always understood to be there. The inverse process of finding the angle given its sines is accomplished by locating the sine value in the table and then reading off the corresponding degrees and minutes.

(c) 0.7290 is located in 46° row and 48′ column, hence the angle is 46° 48′ *Ans* [sin (46° 48′) = 0.7290]

(d) The nearest we can get to 0.8254 is 0.8251 located in the 55° row, 36′ column. We need an extra '3', i.e.

0.8254 = 0.8251 + 0.0003 and this corresponds to the 2′

mean difference column (far right). Hence the angle is

$$55°36' + 2' = 55°38' \quad \textit{Ans}$$

2 Using the portion of the cosine table given below

NATURAL COSINES — SUBTRACT

	0′	6′	12′	18′	24′	30′	36′	42′	48′	54′	1′	2′	3′	4′	5′
45°	.7071	7059	7046	7034	7022	7009	6997	6984	6972	6959	2	4	6	8	10
46	.6947	6934	6921	6909	6896	6884	6871	6858	6845	6833	2	4	6	8	11
47	.6820	6807	6794	6782	6769	6756	6743	6730	6717	6704	2	4	6	9	11
48	.6691	6678	6665	6652	6639	6626	6613	6600	6587	6574	2	4	7	9	11
49	.6561	6547	6534	6521	6508	6494	6481	6468	6455	6441	2	4	7	9	11
50	.6428	6414	6401	6388	6374	6361	6347	6334	6320	6307	2	4	7	9	11
51	.6293	6280	6266	6252	6239	6225	6211	6198	6184	6170	2	5	7	9	11
52	.6157	6143	6129	6115	6101	6088	6074	6060	6046	6032	2	5	7	9	12
53	.6018	6004	5990	5976	5962	5948	5934	5920	5906	5892	2	5	7	9	12
54	.5878	5864	5850	5835	5821	5807	5793	5779	5764	5750	2	5	7	9	12
55	.5736	5721	5707	5693	5678	5664	5650	5635	5621	5606	2	5	7	10	12

NATURAL TANGENTS

	0′	6′	12′	18′	24′	30′	36′	42′	48′	54′	1′	2′	3′	4′	5′
0°	0.0000	0017	0035	0052	0070	0087	0105	0122	0140	0157	3	6	9	12	15
1	0.0175	0192	0209	0227	0244	0262	0279	0297	0314	0332	3	6	9	12	15
2	0.0349	0367	0384	0402	0419	0437	0454	0472	0489	0507	3	6	9	12	15
3	0.0524	0542	0559	0577	0594	0612	0629	0647	0664	0682	3	6	9	12	15
4	0.0699	0717	0734	0752	0769	0787	0805	0822	0840	0857	3	6	9	12	15
5	0.9875	0892	0910	0928	0945	0963	0981	0998	1016	1033	3	6	9	12	15
6	0.1051	1069	1086	1104	1122	1139	1157	1175	1192	1210	3	6	9	12	15
7	0.1228	1246	1263	1281	1299	1317	1334	1352	1370	1388	3	6	9	12	15
8	0.1405	1423	1441	1459	1477	1495	1512	1530	1548	1566	3	6	9	12	15
9	0.1584	1602	1620	1638	1655	1673	1691	1709	1727	1745	3	6	9	12	15
10	0.1763	1781	1799	1817	1835	1853	1871	1890	1908	1926	3	6	9	12	15

find the values of (a) $\cos(45^\circ 18')$ (b) $\cos(49^\circ 21')$

Cosine is normally abbreviated to cos.

Note: You must always subtract the difference in the far right minute column when determining cosines.

Answers

(a) $\cos(45^\circ 18') = 0.7034$

(b) $\cos(49^\circ 21') = 0.6521$ corresponding to $49^\circ 18'$

$-\ 7$ for the extra $3'$

$0.6514 = \cos(49^\circ 21')$

3 Using the portion of the 4-figure tangent value table given on previous page 31, determine

(a) $\tan(8^\circ 13')$, tangent is normally abbreviated to tan,

(b) the angle whose tangent is 0.1125.

Answers

(a) $\tan(8^\circ 13') = 0.1441$ corresponding to $8^\circ 12'$

3 for the extra $1'$

$0.1444 = \tan(8^\circ 13')$

(b) 0.1122 is located 6° row, $24'$ column

3 corresponds to $1'$ in far column

0.1125 therefore corresponds to $6^\circ 24' + 1'$

i.e. angle whose tangent is 0.1125 is $6^\circ 25'$

4.6 The use of conversion tables and charts

Tabular displays, charts, graphical and pictorial displays are widely used as an easy and quick way to find information. Many 'everyday' examples spring to mind: time-tables, interest and loan repayment tables, conversion tables, tax tables – and no doubt you can think of many more.

The following examples show some practical uses of tables and charts and explain how information can be gained from them.

Examples

1 The table displayed in Figure 4.2 shows how your money would grow if you were to make regular monthly savings. In this particular case the table refers to a fixed interest rate of 8.75% over the entire of the savings plan.

The tables are really self-explanatory. We use them in exactly the same way as we did for finding mathematical values. For example, if we wished to invest £100 per month for a period of 5 years we find the value of the investment by simply looking along the £100 row and down the 5 year column: the result £7517.26.

For larger or smaller amounts not listed in the table we could use a little commonsense and also find the result. For example, if we could afford to save £25 per month for 3 years, we could use the table to find the accumulated value of the investment in the following way:

£5 for 3 years yields £206.00
£20 for 3 years yields £824.00
so £25 for 3 years £1030.00

2 Figure 4.3 shows two conversion charts, one for converting millimetres to inches and vice-versa, the other for converting kilograms to pounds and vice-versa.

Those charts provide a straightforward means for converting quantities expressed in one unit to the other. For example:

To convert 8 mm to inches

Use chart Figure 4.3(a):

locate 8 mm on top mm scale, shown as point A; the corresponding inch value is read directly from the bottom inch scale, point B, which gives

8 mm = 0.315

To convert 6.7 kg to lb

Use chart Figure 4.3 (b):

locate 6.7 kg on kg scale, point C;
read off corresponding lb value on lb scale, point D

6.7 kg = 14.8 lb

Regular Savings			
MONTHLY SAVINGS PLAN 8.75% NET INTEREST			
Monthly Savings	Value of investment, assuming continuance of current rate of interest, at the end of a period of:		
	1 YEAR	3 YEARS	5 YEARS
£	£	£	£
5	62.88	206.00	375.86
10	125.75	412.00	751.73
20	251.51	824.00	1,503.45
50	628.77	2,060.01	3,758.63
100	1,257.54	4,120.02	7,517.26
200	2,515.09	8,240.03	15,034.52
500	6,287.72	20,600.08	37,586.30

Figure 4.2 Table showing returns on a regular monthly savings scheme.

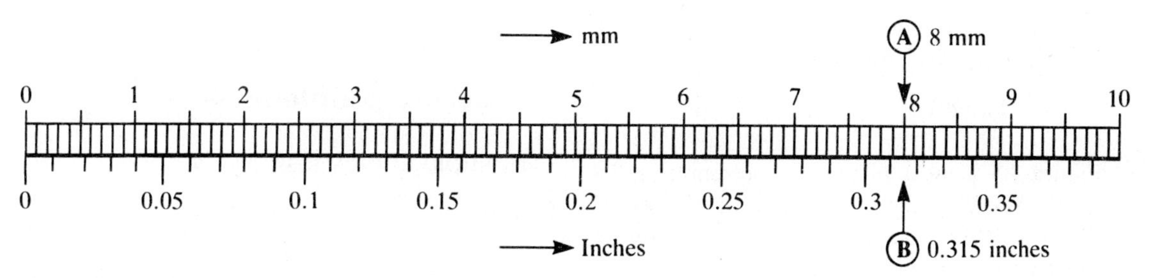

(a) Millimetres to inches and vice-versa

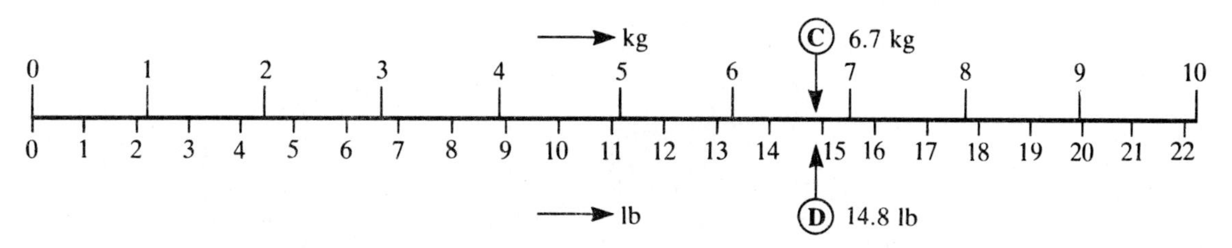

(b) Kilograms to pounds and vice-versa

Figure 4.3 Examples of adjacent-scale conversion charts.

3 The previous example used adjacent scale conversion charts. In this example, we use a parallel scale conversion chart to convert °C (degrees Celsius) to °F (degrees Fahrenheit) and vice-versa. The chart is shown in Figure 4.4.

In this form of chart we use the point *P* as a pivot point. For example, to convert 20°C to°F place a ruler with its straight edge intersecting the 20°C value on the °C scale, point A, and passing through the pivot point P. Read off the

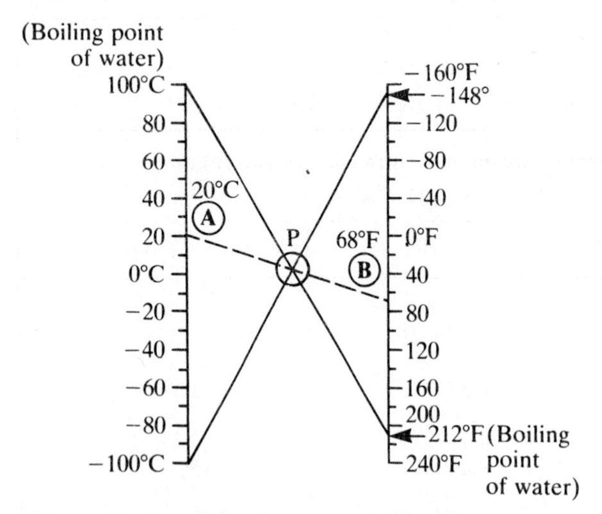

Figure 4.4 Parallel-scale conversion chart for converting temperatures °C to °F and vice-versa.

point where the edge cuts the °F scale, point B. This provides the temperature in °F,

$$20°\text{C} = 68°\text{F}$$

4 Figure 4.5 gives a pictorial representation – a chart – showing the probability or chance of throwing a given score with two dice. Use the chart to find the probability of:

(a) Throwing two sixes.
(b) Making a score with the two dice equal to 9.
(c) Making a score of 4 or below, i.e. 4, 3 or 2.

Solution

(a) The probability of throwing two sixes, i.e. a score of 12, is shown on the chart (far right) and is $\frac{1}{36}$, i.e. one chance in 36.

(b) The chart shows that the probability of throwing a score of 9 is $\frac{1}{9}$ and that this may be achieved in four ways: a 3 and a 6, 4 and 5, 5 and 4, 6 and 3.

(c) The probability of throwing a score of 4 or less is found from the chart by adding the probabilities of throwing scores of 4, 3 and 2, i.e.

$$\text{probability} = \frac{1}{12} + \frac{1}{18} + \frac{1}{36} = \frac{3+2+1}{36} = \frac{6}{36} = \frac{1}{6}$$

Test and problems 4

Multiple choice test: MT 4

Answer block:

Question No.	0	1	2	3	4	5	6	7	8
Answer	b								

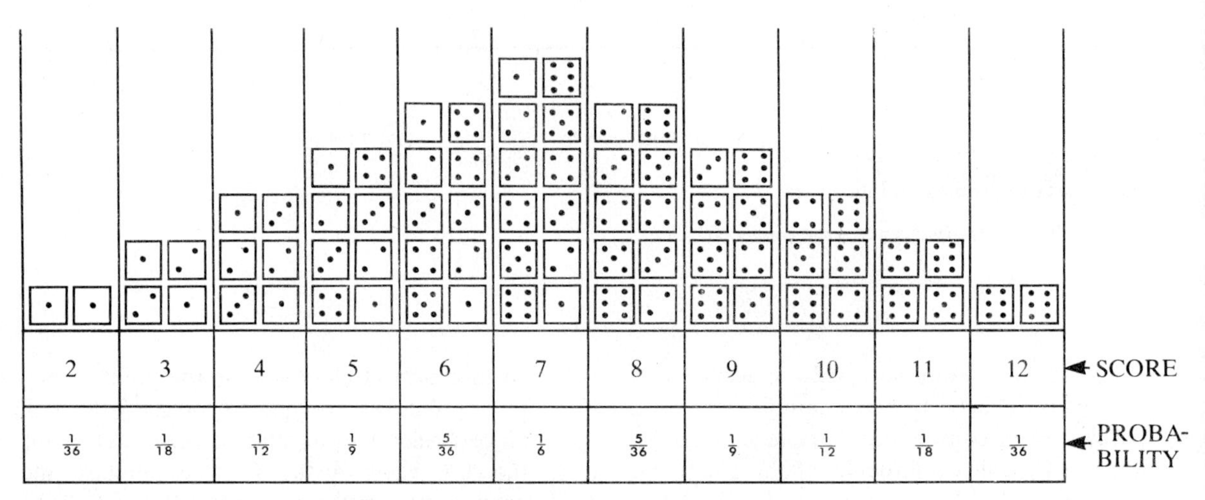

Figure 4.5 Chart showing probabilities of throwing a given score with two dice.

Enter your answer, that is a, b, c or d in the column under the question number in the answer block above. Note that question Qu. O has already been worked out and the answer inserted.

Qu. 0 Using 4-figure log tables evaluate $172.8 \div 24.9$

Ans (a) 6.908 (b) 6.939
(c) 69.40 (d) 7.167

Solution

No.	Log	
172.8	2.2375	subtract to divide
24.9	1.3962	
6.939	0.8413	antilog

Hence the answer is 6.939 and we insert 'b' in the answer block as shown

---

Now carry on with the test.

Qu. 1 Using 4-figure square root tables evaluate $\sqrt{0.0926}$

Ans (a) 0.3043 (b) 0.9623
(c) 0.0304 (d) 0.0962

Qu. 2 Using 4-figure reciprocal tables find the reciprocal of 897.4

Ans (a) 1.14×10^{-3} (b) 111.40
(c) 0.01079 (d) 0.001114

Qu. 3 Using 4-figures tables find the antilog of 4.6276

Ans (a) 4424 (b) 42 422
(c) 4236 (d) 4.2×10^4

Qu. 4 Evaluate using log tables 761.2×0.03676

Ans (a) 27.98 (b) 280.0
(c) 2142 (d) 2978

Qu. 5 Evaluate using log tables $(2.634)^5$

Ans (a) 126.8 (b) 73.83
(c) 69.36 (d) 48.14

Qu. 6 Evaluate by means of log table $(10)^{\frac{1}{6}}$ or equivalently $\sqrt[6]{(10)}$

Ans (a) 1.667 (b) 6.000
(c) 1.468 (d) 1.196

Qu. 7 Determine, using the table of Figure 4.2 the value of investment at the end of a 5-year period when £15 is saved regularly each month. The interest rate of 8.75% may be assumed.

Ans (a) £751.73 (b) £375.86
(c) £978.75 (d) £1127.59

Qu. 8 Using the parallel conversion chart of Figure 4.4, convert $-40°$C to °F.

Ans (a) 40°F (b) $-40°$F
(c) 120°F (d) out of range

Problems 4

1 Using 4-figure squares tables find the squares of
(a) 6.946 (b) 505.7
(c) 0.0462 (d) 7.43×10^{-3}

2 Using 4-figure square root tables find the square roots of
(a) 26.28 (b) 0.7676
(c) 4.8×10^6 (d) 2.67×10^{-3}

3 Using 4-figure reciprocal tables, evaluate
(a) 3.142^{-1} (b) 567^{-1}
(c) 0.142^{-1} (d) $(1.2 \times 10^{-6})^{-1}$

4 Use 4-figure log tables to find the common logarithms of
(a) 6.28; 628; 0.006 28
(b) 4000; 4 000 000; 4×10^{-6}
(c) 0.67; 0.0067; 0.000 067

5 Use 4-figure antilog tables to find the antilogarithms of
(a) 0.6728, $\bar{3}$.6728, 4.6728
(b) 1.4247, $\bar{2}$.4247, 6.4247

6 Evaluate using log tables
(a) 56.23×3.654 (b) $67.29 \div 46.23$
(c) $(87.9 \times 4.44) - (62.6 \times 3.23)$
(d) $(14.62 \div 3.24) + (7.6 \div 0.47)$
(e) 4.627^3 (f) $56.27^{\frac{1}{2}}$ (g) 0.7238^{-1}
(h) $\dfrac{73.2 \times 56.7 \times 39.6}{576.2 - 199.8}$
(i) $\sqrt{\left[\dfrac{6.23 - (0.67 \times 2.43)}{5.26 + 11.47}\right]}$

7 Using the conversion charts of Figures 4.3 and 4.4 convert
(a) 4.5 mm to inches
(b) 0.3 inches to millimetres
(c) 8.1 kg to lb
(d) 5.5 lb to kilograms
(e) 27°C to °F
(f) $-32°$F to °C

8 Using the chart of Figure 4.5 determine
   (a) the probability of throwing the two dice and achieving a score of exactly 8,
   (b) the probability of achieving a score of 9, 10, 11 or 12,
   (c) the probability of throwing a score of 8 or less.

5 Using an electronic calculator

General learning objectives: to perform basic arithmetic operations using a calculator.

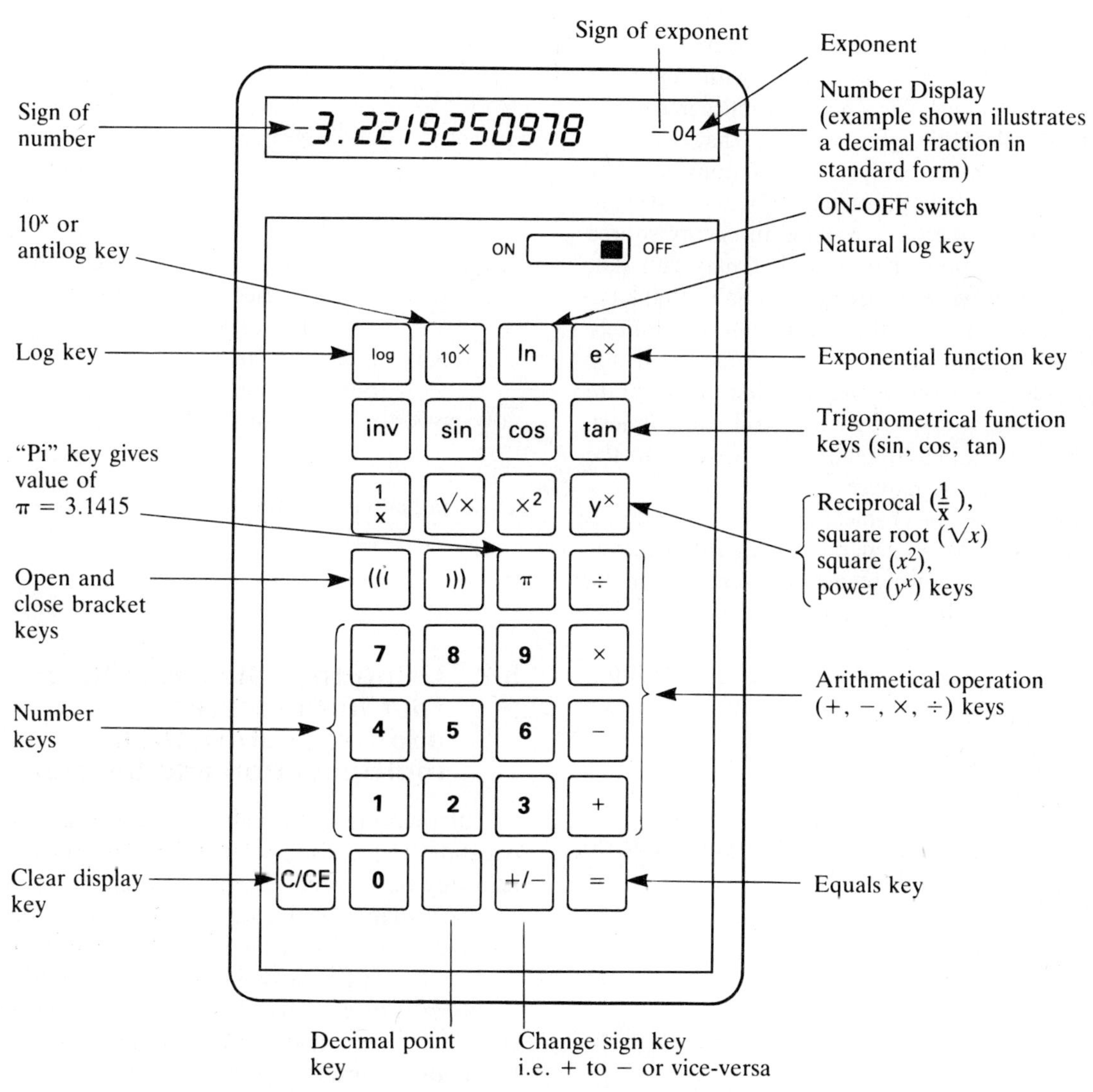

Figure 5.1 Keyboard for a typical electronic calculator suitable for general and scientific use.

5.1 Introduction: use of electronic calculators and typical keyboard layout

Electronic calculators are simple to use and provide very accurate results of typically eight or more significant figures. They have taken the drudgery out of routine calculations and calculations that may take several minutes using 4-figure tables can be worked out in seconds.

A good-quality calculator is relatively inexpensive and is an excellent and strongly recommended investment. It may be used to work out most, if not all, scientific, engineering and business-type calculations likely to be encountered. It also provides a number of important mathematical functions, which may be obtained literally at the 'touch' of the appropriate keys on the front panel.

Once you have become familiar with keyboard layout and have had a few hours practice in operating your calculator, you will find that you can perform even quite complex calculations speedily and reliably. Perhaps the only drawbacks with using calculators are that they require an energy source for their operation – batteries that may run flat, although many versions are now available that have solar cells and thus these can run indefinitely provided a light source is available – and for many practical purposes they provide too accurate answers. The latter 'drawback' is said really in jest, the operator can easily read off the display to the accuracy she or he requires.

Figure 5.1 shows a typical keyboard layout of a calculator suitable for general and scientific calculation work. The following functions, indicated on the face of the individual keys, are available:

[+] add key i.e. press this key to perform addition

[−] subtract key

[×] multiplication key

[÷] divide key

[=] equals key press this key to display answer of a calculation

[0] [1] [2] [3] [4] ··· [9] number keys

[·] decimal point key

[C/CE] clear display key pressed when a calculation has been completed and returns display to 0 so a new calculation can be started; it can also be used to clear last entry if an error has been made in that entry

[((] [))] open and close bracket keys

[π] pi key, gives $\pi = 3.141\,592\,6\cdots$ automatically when pressed

[$\frac{1}{x}$] reciprocal key gives reciprocal of number entered when pressed

[$\sqrt{x}$] square root key

[x^2] square key

[y^x] power key

[sin] [cos] [tan] trigonometrical keys giving sine, cosine and tangent of an angle entered

[inv] inverse trig key used to find angles of given sin, cos and tan values

[log] log key gives the logarithm to base 10 of number entered (normal logs)

[10^x] power of 10 key (antilog key)

[ln] natural log key (log to base e, $e = 2.71828183$)

[e^x] exponential function key

5.2 Using the calculator for the four basic operations: addition, subtraction, multiplication and division

Calculations involving addition, subtraction, multiplication and division of numbers are illustrated below by way of some examples, using an *algebraic logic* calculator. Calculators of this type utilize a straightforward sequence of entry, virtually identical to the order we write down the calculation on paper. For example, in an algebraic logic calculator, numbers and operations are performed in the following order:

$3 + 2$

[3] [+] [2] [=] 5

↑ press 3 key ↑ press + key ↑ press 2 key ↑ press = key ↑ answer displayed

$5 - 2$

[5] [−] [2] [=] 3.

6×4

[6] [×] [4] [=] 24.

$9 \div 5$

[9] [÷] [5] [=] 1.8

The majority of calculators work in this way.

Some calculators, however, employ a different sequence known as *reverse Polish logic*. In reverse Polish logic calculators there is an enter key, ENTER , but **no equals key**. The order of entries for these types of calculators is, for example,

$5 \div 8$

[5] [ENTER] [8] [÷] .625

answer is given in display directly the ÷ key pressed

Algebraic calculators are in general easier to use and for this reason are more commonly recommended for routine calculation work. All the examples given below are for algebraic logic calculators.

Examples

1 Evaluate $17.6 + 4.8$.

Sequence of operations

Switch on calculator. These should be a 0. or 0.000··· form of display, which shows that the calculator battery (or power supply) is operating correctly.

First enter 17.6 by pressing the following keys [1] [7] [·] [6]
Next press [+] key
Then enter 4.8 by pressing [4] [·] [8]
Press [=] key, answer 22.4 will displayed.

Actual step-by-step sequence on calculator:

Key pressed	Number displayed
at switch on	0.
[1]	1.
[7]	17.
[·]	17.
[6]	17.6
[+]	17.6
[4]	4.
[·]	4.
[8]	4.8
[=]	22.4 ← *Answer*

2 Evaluate $0.78 + 0.023 + 4.2$.

Sequence of operations after switching on

Enter 0.78 by pressing [·] [7] [8] (Note: it is not necessary to enter the 0 in front of the decimal point)
Press [+]
Enter 0.023 by pressing [·] [0] [2] [3]
Press [+]
Enter 4.2 by pressing [4] [·] [2]
Press [=], answer 5.003 will be displayed.

3 Evaluate $52.68 - 37.6$.

Sequence

Enter 52.68 by pressing [5] [2] [·] [6] [8]
Press [−]
Enter 37.6 by pressing [3] [7] [·] [6]
Press [=], answer 15.08 will be displayed.

4 Evaluate $2.724 - 412.3 - 1\,047.6 + 512.62$.

Sequence

Enter 2.724 by pressing [2] [·] [7] [2] [4]
Press [−]
Enter 412.3 by pressing [4] [1] [2] [·] [3]
Press [−]
Enter 1 047.6 by pressing [1] [0] [4] [7] [·] [6]
Press [+]
Enter 512.62 by pressing [5] [1] [2] [·] [6] [2]
Press, [=] , answer −944.556 will be displayed.

5 Evaluate 9.26×0.078.

Sequence

Enter 9.26 by pressing [9] [·] [2] [6]

Press [×]

Enter 0.078 by pressing [·] [0] [7] [8]

Press [=], answer 0.72228 will displayed.

6 Evaluate $0.0026 \times 0.000\,52$.

Sequence

Enter 0.0026 by pressing [·] [0] [0] [2] [6]

Press [×]

Enter 0.000 52 by pressing [·] [0] [0] [0] [5] [2]

Press [=], answer 0.000 001 352 or 1.352^{-06}

Note: if display expresses answer in standard-type form:

1.352^{-06} means 1.352×10^{-6}

mantissa index

7 Evaluate $19.2 \div 2\,116$

Sequence

Enter 19.2 by pressing [1] [9] [·] [2]

Press [÷]

Enter 2 116 by pressing [2] [1] [1] [6]

Press [=], answer 9.073724008^{-03} will be displayed on a '10 significant figure' calculator. In most cases it is a waste of time to state the result to such a high degree of accuracy. Four significant figures is usually more than enough. Thus to four significant figures:

$$19.2 \div 2\,116 = 9.074 \times 10^{-3}$$

5.3 Using a calculator to determine numbers to a power, roots, reciprocals and other mathematical functions

As well as the basic arithmetic operation keys, most calculators have a number of mathematical function programs stored in their computer memory. We can utilize these directly to find squares, square roots, other powers and roots, reciprocals, logs, 10^x (antilogs), sines, cosines, tangents and usually many other important functions.

Such is the power in modern-day electronics that it is now possible to pack into a pocket-calculator immense computing power.

To obtain these mathematical function values we enter the required number and just press a key – much easier and far more accurate than tables.

Examples

1 Using a calculator evaluate $\sqrt{50.08}$

Sequence (assuming the calculator has a $\sqrt{x}$ key)

Enter 50.08

Press [$\sqrt{x}$] key

This executes the square root operation and gives the answer directly on the display:

7.076722405

2. Find the reciprocal of 10.636

Sequence

Enter 10.636

Press [1/x] key

This executes the reciprocal operation (divides the number into 1) and gives the answer directly on the display:

9.402030839^{-02} (standard form type of display)

or 0.09402030839 (normal decimal type display)

Alternatively, if there is no [1/x] function key available:

Enter 1

Press [÷]

Enter 10.636

Press [=]

9.402030839^{-02} ...reciprocal displayed.

3 Evaluate 3.67^2.

Sequence (assuming calculator has an [x^2] key)

Enter 3.67

Press [x^2] key

This executes the square operation and gives the answer directly on the display:

13.4689

so $3.67^2 = 13.4689$

4 Evaluate 15.5^3.

Sequence (assuming calculator has $\boxed{y^x}$ key)

Enter 15.5 (the base number, y)

Press $\boxed{y^x}$ key

Enter 3 (the power index x)

Press $\boxed{=}$ to obtain answer 3723.875

so $15.5^3 = 3724$ (to four significant figures)

5 Evaluate 2^{-7}.

Sequence

Enter 2

Press $\boxed{y^x}$

Enter 7

Press $\boxed{+/-}$ this key changes sign from + to −, i.e. index 7 becomes −7.

Press $\boxed{=}$ to obtain answer 7.8125^{-03}

so $2^{-7} = 7.8125 \times 10^{-3}$ or 0.0078125

6 Evaluate $\sqrt[7]{57.64}$

Many calculators have a root function key $\sqrt[x]{y}$. Using this key, the calculation is simply:

Sequence

Enter 57.64

Press $\boxed{\sqrt[x]{y}}$ key

Enter 7

Press $\boxed{=}$

1.784 563 46 ⋯ answer displayed.

7 Using an electronic calculator determine the following trigonometric values:

(a) sin 65° (b) cos 36.2° (c) tan 14°16′

(a) *Sequence for* sin 65°

Type in 65

Press $\boxed{\sin}$

0.906 307 787 is displayed

(b) *Sequence for* cos 36.2°

Type in 36.2

Press $\boxed{\cos}$ cos 36.2° = 0.806 960 312 is displayed.

(c) *Sequence for* tan 14°16′

First we must convert 16′ into a decimal fraction of a degree, i.e. divide 16 by 60, so

Enter 16

Press $\boxed{\div}$

Enter 60

Press $\boxed{+}$

Enter $\boxed{14}$

Press $\boxed{=}$ ⋯ gives 14.266 666 7 (14°16′)

Press $\boxed{\tan}$ ⋯ gives tan(14°16′) = 0.254 277 32

8 Find the angle whose sine is 0.6273

Sequence

Enter 0.6273

Press $\boxed{\text{inv}}$, the inverse function key*

Press $\boxed{\sin}$

38.851 201 16 is displayed, i.e.
sin 38.851 201 16° = 0.6273

*This key may also be called 'arc'. Alternatively, many calculators have a second function key, normally labelled F or 2nd F. When this key is pressed the function of the keys change to provide a second set of functions, usually defined by the printing directly above the keys, as shown, for example in Figure 5.2. $\sin^{-1}$, $\cos^{-1}$, $\tan^{-1}$ are the inverse trigonometric function symbols. They are used to find the angle whose sine, cosine or tangent value is given.

So, for example, to find $\sin^{-1} 0.2354$ (the angle whose sine is 0.2354) using a calculator with the second function key and $\sin^{-1}$ function available:

Enter 0.2354

Press $\boxed{2^{nd}\text{F}}$

Press $\sin^{-1}$ $\boxed{\sin}$

13.615 202 57 is displayed, i.e. $\sin^{-1} = 13.62°$ (to 2 decimal places)

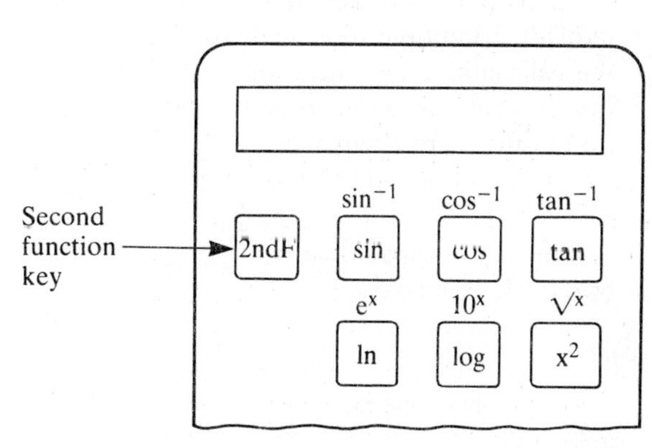

Figure 5.2 Calculator with second function key allowing the availability of more functions.

9 Using a calculator determine
(a) log 257.62 (b) antilog 3.6276 ($10^{3.6276}$)

Sequence

The log and antilog (10^x) functions are very often on the same key, as shown, for example in Figure 5.2. The log function is the 'first function', the antilog is the 'second function' and is brought into operation when required by first pressing the second function key [2ndF.]

Enter 257.62

Press [log]

2.410 979 576 ··· log 257.62 is displayed

Press Clear button, to clear display

We can now proceed to find the antilog or 10^x value:

Enter 3.6276

Press [2ndF], Press [log] (10^x)

4242.286 549 ··· antilog 3.6276 is displayed.

5.4 Using the open and close brackets keys in calculation work

The brackets keys [((( ] and [ )))] are extremely useful. They allow calculations to be performed in a continuous sequence. Their use is illustrated in the following calculations.

Examples

1 Evaluate $7.59 \div 4.32 - 2.67 \times 0.38$.

We must put in brackets in order to group the individual multiplication and division part of the calculation. The inclusion of the brackets instructs the calculator to perform the calculation in the correct sequence.

$(7.59 \div 4.32) - (2.67 \times 0.38)$

open brackets key (first "("), close brackets key (first ")"), open (second "("), close (second ")")

Sequence

Press open brackets key [((( ]

Enter 7.59

Press [÷]

Enter 4.32

Press close brackets key [ )))] ··· result of $(7.59 \div 4.32)$ will now be displayed.

Press [−]

Press open brackets key [((( ]

Enter 2.67

Press [×]

Enter 0.38

Press close brackets key [ )))] ··· result of (2.67×0.38) will be displayed

Press [=] to obtain answer 0.742344444

Answer is 0.7423 to 4 significant figures.

2 Evaluate $(3.97 \times 5.67)^2 - (7.67 \times 2.43)^2$

Sequence

Press open brackets key [((( ]

Enter 3.97

Press [×]

Enter 5.67

Press close brackets key [ )))]

Press [x^2] key ··· result of $(3.97 \times 5.67)^2$ will be displayed

Press [−]

Press open brackets key [((( ]

Enter 7.67

Press [×]

Enter 2.43

Press close brackets key [ )))]

Press [x^2] key ··· result of $(7.67 \times 2.43)^2$ will be displayed

Press [=] to obtain answer 159.3168264

Answer is 159.32 to 2 decimal places.

3 Evaluate

$$\frac{5.6 \times 2.3 - 7.2 \times 1.9}{10.2 \times 4.37 + 9.4 \times 3.74}$$

Here we put in brackets to group the individual parts in both numerator and denominator, and then a second time to group the complete numerator and complete denominator, i.e.

$$N = (5.6 \times 2.3) - (7.2 \times 1.9)$$
$$D = (10.2 \times 4.37) + (9.4 \times 3.74)$$
$$\frac{N}{D} = N \div D = (N) \div (D)$$

so we write our calculation in the form

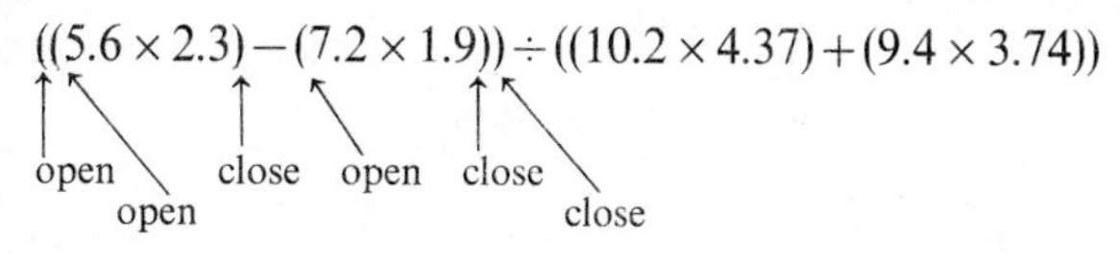

Sequence	*Display*
Press open brackets key [((]	0.
Press [((] a second time	0.
Enter 5.6	5.6
Press [×]	5.6
Enter 2.3	2.3
Press close brackets key [))]	12.88
Press [−]	12.88
Press [((]	0.0
Enter 7.2	7.2
Press [×]	7.2
Enter 1.9	1.9
Press close brackets key [))]	13.68
Press [))] a second time	−0.8
Press [÷]	−0.8
Press [((]	0.
Press [((]	0.
Enter 10.2	10.2
Press [×]	10.2
Enter 4.37	4.37
Press [))]	44.574
Press [+]	44.574
Press [((]	0.
Enter 9.4	9.4
Press [×]	9.4
Enter 3.74	3.74
Press [))]	35.156
Press [))]	79.73

Press [=] to obtain answer -1.003386429^{-02}
Answer to four significant figures is -1.003×10^{-2}.

5.5 Checking calculations and results

Although calculators are extremely accurate, the human operator may inadvertently produce errors. He or she may enter wrong values, not fully understand the function of some of the keys and/or present the calculator with the incorrect series of operations to effect the given calculation.

So, as already explained in Chapter 3, it is good practice to make a rough calculation to serve as a check on the results obtained using a calculator.

Examples

1 Evaluate

$$\frac{36\,756.7 \times 56.2 \times 3600.9}{0.62 \times 7.28 \times 0.056}$$

using a calculator but first make a rough calculation to find an approximate result.

Rough calculation

Expressing the individual terms in standard form rounded to 1 or 2 significant figures and then employing rough cancelling, we have:

$$\frac{\overset{0.6}{\cancel{3.7}} \times 10^4 \times \overset{1}{\cancel{5.6}} \times 10^1 \times \overset{0.5}{\cancel{3.6}} \times 10^3}{\underset{1}{\cancel{6}} \times 10^{-1} \times \underset{1}{\cancel{7}} \times \underset{1}{\cancel{5.6}} \times 10^{-2}}$$

$$\approx \frac{0.6 \times 0.5 \times 10^{4+1+3}}{10^{-1-2}} = 0.30 \times 10^{8-(-3)}$$

$$= 0.3 \times 10^{11} = 3 \times 10^{10}$$

By calculator:

Press [((] and enter 36 756.7 [×] 56.2 [×] 3600.9 [))] ··· this evaluates numerator, yielding 7438 474 698. Press [÷], [((] .62 [×] 7.28 [×] 0.056 [))] ··· this evaluates denominator, yielding 0.252 761 6 Press = to give answer: $2.942\,881\,6 \times 10^{10}$. Our rough and calculator results check well, so with confidence we can state the result:

2.9429×10^{10} to 4 decimal places.

2 Evaluate: $\left(\frac{1}{9.6} + \frac{1}{20.8}\right)^{-1}$

Rough calculation

$$\left(\frac{1}{9.6}+\frac{1}{20.8}\right)\approx\left(\frac{1}{10}+\frac{1}{20}\right)=\frac{2+1}{20}=\frac{3}{20}$$

$$\left(\frac{1}{9.6}+\frac{1}{20.8}\right)^{-1}\approx\left(\frac{3}{20}\right)^{-1}=1\div\frac{3}{20}$$

$$=1\times\frac{20}{3}\approx 6.7$$

By calculator:

Press [((( ], enter 9.6, press [$1/x$] key, press [ )))] ··· determines $\frac{1}{9.6}$

Press [+]

Press [((( ], enter 20.8, press [$1/x$] key, press [ )))] ··· determines $\frac{1}{20.8}$

Press [=] ··· display gives $\frac{1}{9.6}+\frac{1}{20.8}=$ 0.152 243 589

Press [$1/x$] ··· gives result, 6.568 421 053

i.e. $\left(\frac{1}{9.6}+\frac{1}{20.8}\right)^{-1}=6.5684$ to 4 decimal places

3 Evaluate $\left(\frac{6.28\times 44.2-3.47\times 21.7}{6.8\times 5.3+47.2\times 0.4}\right)^2$

Rough calculation

$$\left(\frac{6\times 40-3\times 20}{7\times 5+50\times .4}\right)^2=\left(\frac{240-60}{35+20}\right)^2$$

$$=\left(\frac{180}{55}\right)^2\approx 3^2=9$$

By calculator: a typical sequence

Press open brackets twice, i.e. [((( ] [((( ]

Enter 6.28 [×] 44.2, press [ )))]

Press [−], press [((( ], enter 3.47 [×] 21.7, press [ )))] [ )))]

Press [÷]

Press [((( ] [((( ], enter 6.8 [×] 5.3, press [ )))]

Press [+], press [((( ], enter 47.2 [×] .4, press [ )))] [ )))]

Press [=] and finally the [x^2] key to find the square and thus the final result.

The answer 13.525 945 37 is displayed. This answer, at first sight, is at variance with our rough calculation of 9. Make a second more accurate check, using the calculator for the step-by-step calculations.

Second rough calculation

$$6.28\times 44.2-3.47\times 21.7\approx 277-75=202$$
$$6.8\times 5.3+47.2\times 0.4\approx 36+19=55$$

Dividing numerator by denominator and squaring:

$$\left(\frac{202}{55}\right)^2\approx(3.67)^2\approx 13.5$$

So in fact our calculation appears correct.

Test and problems 5

Multiple choice test: MT 5

Answer block:

Question No.	0	1	2	3	4	5	6	7	8
Answer	b								

Enter your answer, that is a, b, c or d in the column under the question number in the answer block above. Note that question Qu. 0 has already been worked out and the answer inserted.

Qu. 0 Evaluate 3.141 59 ÷ 2.718 28 with the aid of a calculator, expressing your answer to 4 decimal places

Ans (a) 1.555 727 15 (b) 1.1557
(c) 1.156 (d) 8.5340

Solution

Enter 3.141 59, press the [÷] key, enter 2.718 28. Press [=] for answer 1.155 727 151

Round to 4 decimal places: 1.1557 *Ans*, so we enter ‘b’ under Qu. 0 in the answer block

---

Now carry on with the test.

Qu. 1 Determine $0.047\,62^{-1}$ to 4 significant figures

Ans (a) 20.9996 (b) 20.99
(c) 21.00 (d) 2.099 958

Qu. 2 Determine $93\,000 \times 186\,000$ expressing your result in standard form to 3 significant figures.

Ans (a) 1.7298×10^{10} (b) 1.73×10^{10}
(c) 0.500 (d) 1.73×10^{6}

Qu. 3 Evaluate $\sqrt{(4.62/0.78)}$ expressing your result to 4 decimal places.

Ans (a) 5.9231 (b) 2.4337
(c) 35.0828 (d) 1.8983

Qu. 4 Use your calculator to estimate the result of the calculation

$$\frac{73.699 \times 1.876 - 2.606 \times 84.7}{114.6 \times 56.2 + 214.7 \times 28.8}$$

to 2 significant figures

Ans (a) 6.5×10^{-3} (b) -140
(c) -0.0065 (d) 140

Qu. 5 Evaluate $27.3 \times 10^{-1.4}$ correct to 4 significant figures

Ans (a) 685.7 (b) 0.9201
(c) 1.086 83 (d) 1.087

Qu. 6 Evaluate $(6.6 \times 3.2)^2 - (6.6 \times 1.2)^2$

Ans (a) 174.24 (b) 383.328
(c) 13.2 (d) 56.34

Qu. 7 Evaluate $4.78 + \sqrt{(16.27 + 5.26 \times 4.7)}$ correct to 2 decimal places.

Ans (a) 11.18 (b) 1.62
(c) 9.75 (d) 45.77

Qu. 8 Evaluate $2.62 \times 10^3 + 6.37 \times 10^4 - 7.67 \times 10^5$

Ans (a) 1.32×10^3 (b) 8.33×10^5
(c) -2.62×10^4 (d) -7.0068×10^5

Problems 5

Evaluate the following using a calculator and quote the results to 4 significant figures:

1 $42.78 + 6.29 + 139.43 + 57.76$
2 $437.06 - 111.28 - 567.26 + 363.67$
3 44.46×56.27
4 $52.3 \times 6.12 - 114.7 \times 0.43$
5 $\dfrac{3.67 \times 4.62 + 5.67 \times 4.22}{0.63 \times 0.28}$
6 5.67^2
7 33.84^{-1}
8 $\log 5.4 + \log 6.28$
9 $3.6^{-1} + 3.6^{-2}$
10 $\sqrt{(6.28 \times 4.34)}$
11 3.2^5
12 6.4^{-3}
13 $10^{3.6}$
14 $\cos 73.24°$
15 $\sin 64° + \cos 26°$

Part Two: Algebra

6 Basic notation and rules of algebra

General learning objectives: to understand and to use algebraic notation and to apply the rules of algebra.

6.1 Introduction: algebra and its use

Algebra is the branch of mathematics which deals with the properties of, and the relationships between, quantities expressed in terms of symbols rather than numbers alone.

An understanding of the use of algebraic notation and the rules of manipulating algebraic expressions provides us with very powerful mathematical tools. For example, it provides general methods for defining and solving problems. Using algebraic notation we can set up equations employing symbols for the 'unknowns' and then solve these equations using the rules of algebra. We can use algebraic notation to express the results in a general manner by means of a formula. Using the rules of algebra we may transform a formula to find a 'wanted' quantity in terms of the others. We can then substitute into the algebraic expression the numerical values of the quantities which are known to obtain the value of the unknown quantity.

In this first chapter on algebra we introduce algebraic notation and then consider the basic rules: the addition, subtraction, multiplication and division of algebraic expressions, and the laws of indices. The rules of manipulating algebraic expressions are in fact identical to those we have already been using for numbers, as of course they must be since in algebra we are essentially only using symbols to represent numerical values of quantities in a generalized way.

6.2 Algebraic notation: the use of symbols and some conventions

The use of symbols – English and sometimes Greek letters – in place of numbers allows general mathematical statements to be written down rather than just specific ones.

By employing symbols to represent numerical quantities we can express physical laws and practical results, formulae, etc. in a generalized way. For example,
Ohm's law (see Figure 6.1a)

$$\text{voltage} = \text{resistance} \times \text{current}$$

can be written as the algebraic equation,

$$V = R \times I$$

where V is the symbol used to represent voltage, units volts
R is the symbol for resistance, units ohms
I is the symbol for current, units amperes

Newton's second law of motion (see Figure 6.1b)

$$\text{force} = \text{mass} \times \text{acceleration}$$

can be expressed as,

$$F = m \times a$$

where the symbols F, m and a are used to denote force, mass and acceleration, respectively.

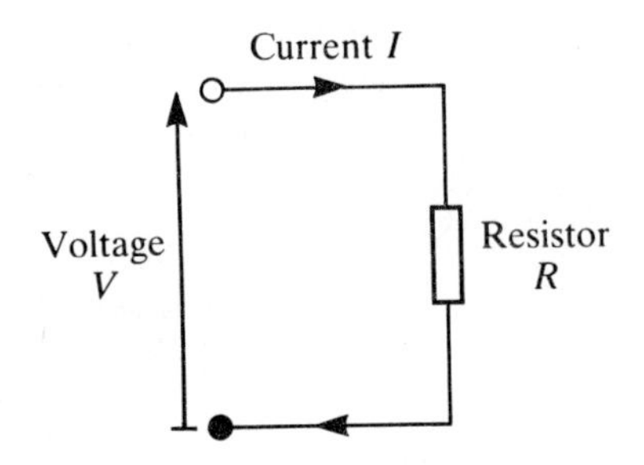

(a) Ohm's law: $V = R \times I$

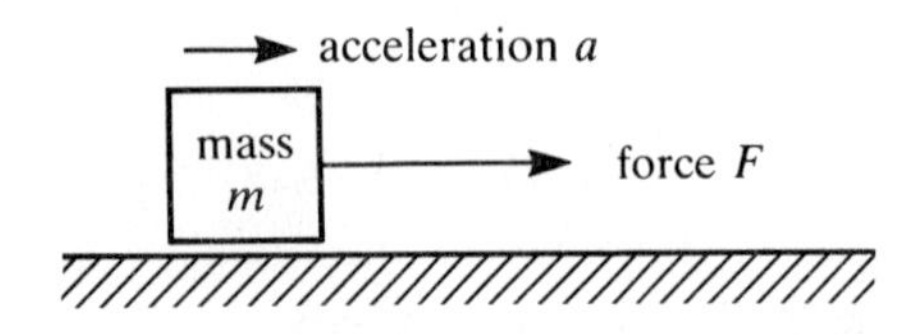

(b) Newton's second law of motion: $F = m \times a$

Figure 6.1.

The simple-interest formula,

$$\text{Interest} = \frac{\text{amount invested} \times \text{time in years} \times \text{rate of interest\%}}{100}$$

can be written in algebraic form as

$$I = \frac{P \times T \times R}{100}$$

where we have used the symbols

P to denote amount invested, the principal
T to denote the time in years
R to denote the interest rate in % per annum
and I to denote the interest gained

The formula for compound interest may be expressed as

$$A = P + I = P(1 + R/100)^T$$

where A = amount invested plus interest accrued
P = amount invested, the principal
I = interest
R = rate of interest in % per annum
T = time in years

In writing down algebraic expressions the multiplication sign is normally omitted, e.g.

$5x$ means $5 \times x$ or 'five' x
xyz means $x \times y \times z$

Ohm's law may be written as $V = RI$

Division is normally, as in the case of numbers, denoted by a horizontal line and sometimes by the solidus /, e.g.

$$\frac{5x - 2y}{9x + 3z} \text{ means } (5x - 2y) \div (9x + 3z)$$

$$(2a + b)/(a + b) \text{ means } (2a + b) \div (a + b)$$

Brackets, again as in the case of numbers, are used to group terms, so for example, if we were to evaluate

$$(2a + b)/(a + b) \quad \text{for the case } a = 3, \quad b = 5$$

the bracket terms must be evaluated first and afterwards the division:

$$(2a + b)/(a + b) = (2 \times \underset{a}{\uparrow}3 + \underset{b}{\uparrow}5)/(\underset{a}{\uparrow}3 + \underset{b}{\uparrow}5)$$

$$= (6 + 5)/8 = 11/8 = 1\tfrac{3}{8}$$

Powers, roots, and reciprocals are written in the identical manner to that used for numbers, e.g.

a^2 means $a \times a$, a squared
$6b^3$ means $6 \times b \times b \times b$, 'six b cubed'
$\sqrt{c}$, $c^{\frac{1}{2}}$ both denote the square root of c
$x^{-1} = \frac{1}{x}$, the reciprocal of x

Examples

1 Evaluate the following algebraic expressions for the case when

$$a = 2, \quad b = 3, \quad c = 5$$

(a) $4ab$ (b) c^{-1} (c) $(a + c)/(c - a)$
(d) $3a(5b - 2c)$

Solution

(a) $4ab = 4 \times \underset{a}{\uparrow}2 \times \underset{b}{\uparrow}3 = 24$

(b) $c^{-1} = \frac{1}{c} = \frac{1}{5}$ or 0.2

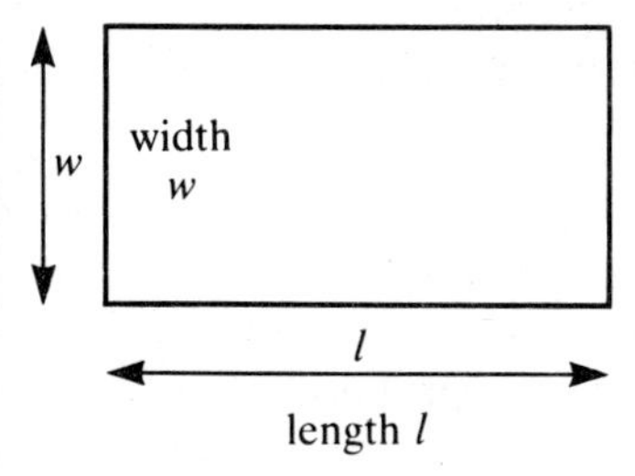

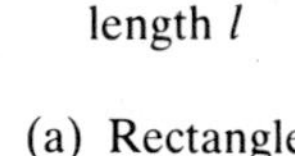
(a) Rectangle

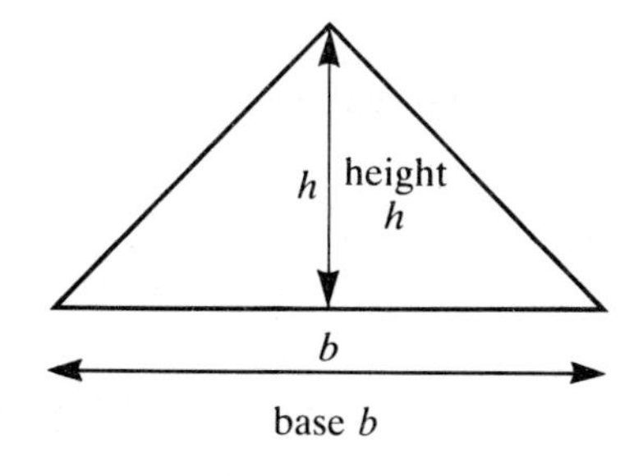

(b) Triangle

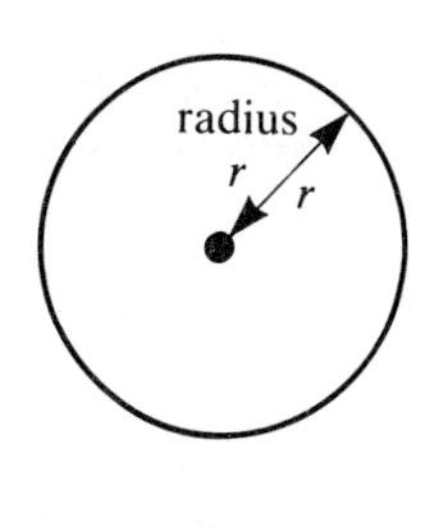

(c) Circle

Figure 6.2.

(c) $\frac{(a+c)}{(c-a)} = \frac{(2+5)}{(5-2)} = \frac{7}{3} = 2\frac{1}{3}$

(d) $3a(5b-2c) = 3 \times a \times (5 \times b - 2 \times c)$
$= 3 \times 2 \times (5 \times 3 - 2 \times 5)$
$= 6 \times (15 - 10) = 6 \times 5 = 30$

2 Ohm's law may be expressed as $V = RI$
Evaluate the voltage V for the case of a resistance of $R = 50$ ohms when carrying a current of 3 amperes

Solution

$$V = RI = 50 \times 3 = 150 \text{ volts}$$

3 Figure 6.2 shows the diagrams of a rectangle, triangle and a circle. Their areas are given by

rectangle: area = width × length
triangle : area = one half × base × height
circle : area = π × radius squared
where π is a constant, equal to 3.142 to 3 decimal places

Using the symbol A to denote area and the symbols for the dimensions as shown in Figure 6.2, express the above results as algebraic formulae.

Solution

(a) Rectangle: $A = wl$
(b) Triangle: $A = \frac{1}{2}bh$
(c) Circle: $A = \pi r^2$

4 Determine the areas of (a) a rectangle of dimension 5.6 by 4.2, (b) a triangle of base length 12.2 and perpendicular height 6.4, (c) a circle of radius 7.6

Answers

(a) $A = wl = 5.6 \times 4.2 = 23.52$
(b) $A = \frac{1}{2}bh = \frac{1}{2} \times 12.2 \times 6.4 = 39.04$
(c) $A = \pi r^2 = 3.142 \times 7.6^2 = 181.48$

5 Using the compound interest formula:

$$A = P + I = P(1 + R/100)^T$$

where the symbols have been previously defined, determine the total amount accrued for an investment of £1000 at 8.5% per annum interest over a period of 10 years.

Solution

$P = 1000, R = 8.5, T = 10$ so on substituting into the formula, we have

$$A = 1000(1 + 8.5/100)^{10}$$
$$= 1000(1 + 0.085)^{10} = 1000 \times 1.085^{10}$$
$$= 1000 \times 2.2609834 = £2,260.98$$

So the £1000 invested has increased to £2,260.98 of which the interest $I = A - £1000 = £1,260.98$

6.3 The laws of algebra: commutative, associative and distributive laws and laws of precedence

Before dealing specifically with the methods of how we add, subtract, multiply and divide algebraic expressions let us first state the basic rules or laws we use in algebra. These laws are basically commonsense rules we have already been using for numbers – since the letters we use in algebra are simply symbols for numerical quantities.

1 **Commutative law**
This law states that additions and subtractions may be performed in any order, e.g.

$$x + y = y + x$$
$$x + y - z = x - z + y$$

and also that multiplication can be performed in any order, e.g.

$$xy = yx$$
$$xyz = zxy = yzx$$

2 **The associative law**
This law states that terms in an algebraic expression may be grouped in any order, e.g.

$$x + y + z = (x + y) + z$$
$$= x + (y + z)$$
$$xyz = x(yz) = (xy)z$$

3 **The distributive law**
This law states that the product of a compound expression and a single term is the algebraic sum of the products of the single term with all terms in the expression, e.g.

$$x(y + z) = xy + xz$$
$$3x(2y - 9z) = 6xy - 27xz$$

4 **Laws of precedence**
These laws or rather rules dictate the order in which we must effect algebraic operations. They are identical to the rules of precedence we have already been using in arithmetic:

FIRST deal with terms in *B*rackets
THEN work out '*O*f', *D*ivision and *M*ultiplication
FINALLY work out *A*dditions and *S*ubtractions

The order of working may be remembered by applying the **BODMAS** rule:

B O D M A S
↑ ↑ ↑ ↑ ↑ ↑
Of ÷ × + −
Brackets
(FIRST) (LAST)

The four above laws together with the laws of indices form the basis of manipulating and simplifying algebraic expressions.

6.4 The addition and subtraction of algebraic expressions

In addition and subtraction of algebraic terms we essentially apply the commutative law (addition and subtraction can be performed in any order) and the associative law (terms may be grouped in any order).

Thus in adding and subtracting algebraic terms to simplify an algebraic expression, we use the following rules:

1 Addition and subtraction can be performed in any order.
2 Like terms, e.g. all x terms, all y terms, all xy, all x^2 terms, etc., may be collected together and combined into a single term; unlike terms cannot be combined, e.g. $7x^2 + 4x + y^2 - 3xy + 49$ cannot be simplified any further.

Examples

1 Simplify $7x - 5x + 3x + 9x - 10x$

Simplification:

First group all positive terms together and all negative terms together, i.e.

$$\underbrace{7x + 3x + 9x}_{= 19x} \underbrace{- 5x - 10x}_{- 15x} = 4x \quad Ans$$

2 Find the sum of $2y^2, 3x^2, 4x, 4x^2, y^2, -x$

Solution

Grouping all like terms together, we have

$$\text{Sum} = \underbrace{2y^2 + y^2}_{3y^2} + \underbrace{3x^2 + 4x^2}_{7x^2} + \underbrace{4x - x}_{3x}$$
$$= 3y^2 + 7x^2 + 3x \quad Ans$$

3 Find the sum of $2x^2 + 5x - 6$ and $7x^2 + 3x - 1$

Solution

This can be accomplished directly by adding like terms or by setting out in columns with like terms under one another, i.e.

$$\begin{array}{l} 2x^2 + 5x - 6 \\ 7x^2 + 3x - 1 \\ \hline 9x^2 + 8x - 7 \quad Ans \end{array}$$

4 Simplify $(4x^3 - 2) + (2x^3 + 4x^2 + 1) + x^3 + 5x^2 + 3x - 3$

Solution

Again this may be accomplished by removing brackets and adding like terms or by setting out as in example 3, in columns. Using the former method, we have

$$\underbrace{4x^3 + 2x^3 + x^3} + \underbrace{4x^2 + 5x^2} + 3x \underbrace{- 2 + 1 - 3}$$
$$= 7x^3 \quad + \quad 9x^2 \quad + 3x - \quad 4$$
$$= 7x^3 + 9x^2 + 3x - 4 \quad Ans$$

5 Evaluate $(2x^2 + 5xy + 6) - (x^2 + 3xy + 4)$

Solution

This may be done by subtracting like terms, i.e.

$$2x^2 - x^2, \quad 5xy - 3xy, \quad 6 - 4$$

so the expression reduces to

$$x^2 + 2xy + 2 \quad Ans$$

or by setting out in columns with like terms under one another and then subtracting, i.e.

$$\begin{array}{r} 2x^2 + 5xy + 6 \\ x^2 + 3xy + 4 \\ \hline x^2 + 2xy + 2 \end{array} \quad Ans$$

6 Subtract $a^3 - 3a^2 + 7$ from $5a^3 + 2a^2 + 7a - 6$

Solution

$$\begin{array}{l} 5a^3 + 2a^2 + 7a - 6 \\ \;\; a^3 - 3a^2 \qquad\;\; + 7 \\ \hline 4a^3 + 5a^2 + 7a - 13 \\ \hline \end{array} \quad Ans$$

Note: $2a^2 - (-3a^2) = 2a^2 + 3a^2 = 5a^2$
$\qquad -6 - (+7) \quad = -6 - 7 \quad = -13$

See rules of signs, section 6.5.

A simple useful rule that can be used for subtraction when setting out in columns is:

to subtract: change sign in all terms in bottom line and add

e.g. $(5a^3 + 2a^2 + 7a - 6) - (a^3 - 3a^2 + 7)$

$$\begin{array}{l} = \quad 5a^3 + 2a^2 + 7a - 6 \\ \;\; + - a^3 + 3a^2 \qquad - 7 \\ \hline \quad 4a^3 + 5a^2 + 7a - 13 \\ \hline \end{array} \quad Ans$$

(signs in bottom line have changed)

6.5 Simple multiplication and division of algebraic terms; rules of signs and laws of indices

Multiplication and division is exactly similar when working with algebraic terms as already used with numbers. However, it is important to remember the following 'rules of signs':

$+a \times +b = +ab$	plus time plus gives plus
$-a \times +b = -ab$	minus times plus or
$+a \times -b = -ab$	vice-versa gives minus
$-a \times -b = +ab$	minus times minus gives plus

and for division:

$$\frac{+a}{+b} = +\frac{a}{b}, \quad \frac{-a}{+b} = -\frac{a}{b},$$

$$\frac{+a}{-b} = -\frac{a}{b}, \quad \frac{-a}{-b} = +\frac{a}{b}$$

Examples

1 $5a \times -4b = 5 \times -4 \times a \times b$
$\qquad = -20 \times ab = -20ab \quad Ans$

Note the procedure:
first multiply the coefficients (numbers in front of a and b), then, the resultant product = (number product) ab

2 $-4x \times -5y \times -3z = (-4 \times -5 \times -3)xyz$
$\qquad = (+20 \times -3)xyz$
$\qquad = -60xyz \quad Ans$

3 $\dfrac{15p}{3q} = \dfrac{\overset{5}{\cancel{15}}p}{\underset{1}{\cancel{3}}q} = \dfrac{5p}{q}$ or $5\dfrac{p}{q}$ $\quad Ans$

4 $\dfrac{64a}{-128b} = -\dfrac{\overset{1}{\cancel{64}}a}{\underset{2}{\cancel{128}}b} = -\dfrac{a}{2b}$ or $-\dfrac{1}{2}\dfrac{a}{b}$ $\quad Ans$

5 $\dfrac{3x}{2y} \times \dfrac{4y}{9x} = \dfrac{\overset{1}{\cancel{3x}}}{\underset{1}{\cancel{2y}}} \times \dfrac{\overset{2}{\cancel{4y}}}{\underset{3}{\cancel{9x}}} = \dfrac{2}{3}$ $\quad Ans$

To accomplish multiplication and division and to simplify algebraic expressions we also apply the laws of indices, previously considered in Chapters 1 and 2. Remember

1 To multiply, add indices: $a^m \times a^n = a^{m+n}$

2 To divide, subtract indices: $\frac{a^m}{a^n} = a^{m-n}$

3 To raise to a power, multiply indices:

$$(a^m)^n = a^{mn}$$

4 To find a root, divide indices: $\sqrt[n]{(a^m)} = a^{m/n}$

5 Meaning of a negative index: $a^{-n} = \frac{1}{a^n}$

6 Any quantity to power 0 equals 1: $a^0 = 1$

Examples

1 Express as powers of x:

(a) $(x^3)^4$ (b) $(2x)^5$

(c) $(-3x)^3$ (d) $(10x^2)^{-1}$

Answers

(a) $(x^3)^4 = x^{3\times4} = x^{12}$

(b) $(2x)^5 = 2^5 \times x^5 = 32x^5$

(c) $(-3x)^3 = -3^3x^3 = -3 \times -3 \times -3x^3$
$= -27x^3$

(d) $(10x^2)^{-1} = \frac{1}{10x^2} = \frac{1}{10} \times x^{-2} = 0.1x^{-2}$

2 Simplify the algebraic expressions

(a) $2xy^2 \times 5x^3y^4$ (b) $25a^2b^3 \div 5a^2b$

(c) $\frac{4x^2y^3}{3xy} \times \frac{15x^2y^2}{16xy}$ (d) $(16x^4)^{\frac{1}{2}} \times (27x^6)^{\frac{1}{3}}$

Answers

(a) $2xy^2 \times 5x^3y^4 = (2\times5)x^{1+3}y^{2+4} = 10x^4y^6$

(b) $\frac{25a^2b^3}{5a^2b} = \frac{\overset{5}{\cancel{25}}\cancel{a^2}b^3}{\underset{1}{\cancel{5}}\cancel{a^2}b} = \frac{5b^3}{b} = 5b^{3-1} = 5b^2$

(c) $\frac{\overset{1}{\cancel{4}}x^2y^3}{\underset{1}{\cancel{3}}xy} \times \frac{\overset{5}{\cancel{15}}x^2y^2}{\underset{4}{\cancel{16}}xy} = \frac{5x^{2+2}y^{3+2}}{4x^{1+1}y^{1+1}}$

$= \frac{5x^4y^5}{4x^2y^2} = \frac{5x^{4-2}y^{5-2}}{4} = \frac{5}{4}x^2y^3$

(d) $(16x^4)^{\frac{1}{2}} = 4x^2$, $(27x^6)^{\frac{1}{3}} = 3x^2$

as $16^{\frac{1}{2}} = 4$, $x^{4\times\frac{1}{2}} = x^2$, $27^{\frac{1}{3}} = 3$, $x^{6\times\frac{1}{3}} = x^2$

so $(16x^4)^{\frac{1}{2}} \times (27x^6)^{\frac{1}{3}} = 4x^2 \times 3x^2 = 12x^4$

6.6 Simplification of algebraic expressions: applications of laws of algebra and laws of precedence

In this introductory chapter we have applied the laws of algebra to add, subtract and to perform simple multiplication and division and thereby reduce algebraic expressions to usually, although not always, a simpler form. To complete this chapter we revise this work and also apply the laws of precedence to simplify algebraic expressions.

Examples

1 Simplify the following algebraic expressions, i.e. remove brackets, carry out arithmetic operations according to the BODMAS rule.

(a) $5x(x+2) - 4x(3x-1) + 3$

(b) $3\{2x^2 - 5x(x+3) + 3(x^2+5x+1)\}$

Simplification

(a) First remove brackets following the distributive rule, i.e. all terms within the brackets must be multiplied by the multiplying term in front of the brackets.

$5x(x+2) - 4x(3x-1) + 3$
$= 5x^2 + 10x - 12x^2 + 4x + 3$
$= \underbrace{5x^2 - 12x^2} + \underbrace{10x + 4x} + \underbrace{3}$
$= -7x^2 \quad + 14x \quad + 3$
$= -7x^2 + 14x + 3$ *Ans*

(b) $3\{2x^2 - 5x(x+3) + 3(x^2+5x+1)\}$
$= 3\{2x^2 - 5x^2 - 15x + 3x^2 + 15x + 3\}$,
i.e. remove inner brackets first
$= 3\{\underbrace{2x^2 - 5x^2 + 3x^2} - \underbrace{15x + 15x} + \underbrace{3}\}$
$= 3\{\quad 0 \quad\quad 0 \quad + 3\}$
$= 3\{3\} = 9$ *Ans*

2 Simplify

(a) $5(2x-3y) - 3[x + 4\{2 - 6(x-2y)\}]$

(b) $\frac{(3ab)^2}{a(2a-b) - 2b(b-a) - 2(a^2-b^2)}$

Simplification

(a) $5(2x-3y) - 3[x+4\{2-6(x-2y)\}]$
$= 10x - 15y - 3[x + 4\{2 - 6x + 12y\}]$
$= 10x - 15y - 3[x + 8 - 24x + 48y]$
$= 10x - 15y - 3x - 24 + 72x - 144y$
$= 10x - 3x + 72x - 15y - 144y - 24$
$= 79x - 159y - 24$ *Ans*

(b) First simplify denominator by removing brackets and collecting like terms, i.e.

denominator $= 2a^2 - ab - 2b^2 + 2ab - 2a^2 + 2b^2$
$= 2a^2 - 2a^2 - ab + 2ab - 2b^2 + 2b^2$
$= ab$

numerator $= (3ab)^2 = 9a^2b^2$

so on dividing by denominator and cancelling:

$$\frac{9a^2b^2}{ab} = 9ab \quad \textit{Ans}$$

Test and problems 6

Multiple choice test: MT6

Answer block:

Question No.	0	1	2	3	4	5	6	7	8
Answer	c								

Enter your answer, that is a, b, c or d in the column under the question number in the answer block above. Note that question Qu 0 has already been worked out and the answer inserted.

Qu. 0 Subtract $3k^2 - 5k + 7$ from $5k^2 + 5k - 7$
Ans (a) $2k^2$ (b) $8k^2$ (c) $2k^2 + 10k - 14$
(d) $2k^2 - 10k + 14$

Solution

$(5k^2 + 5k - 7) - (3k^2 - 5k + 7)$
$= 5k^2 + 5k - 7 - 3k^2 + 5k - 7$
$= 5k^2 - 3k^2 + 5k + 5k - 7 - 7$
$= 2k^2 + 10k - 14$ *Ans*

so the correct answer is (c) and we insert 'c' under question Qu No. 0 in the answer block

---

Now carry on with the test.

Qu. 1 Add together the terms

$(7x + 3)$, $(3x + 2)$, $x^2 + 7x + 11$

Ans (a) $x^2 + 14x + 16$ (b) $x^2 + 17x + 16$
(c) $x^2 + 3x + 6$ (d) $17x + 16$

Qu. 2 Given $p = 100$, $q = 12$, $r = 3$ evaluate

$\frac{1}{3}pq^2r$

Ans (a) 1200 (b) 3600
(c) 4000 (d) 14 400

Qu. 3 Divide $12x^3b^2c$ by $3x^2b$
Ans (a) $48x^5b^3c$ (b) $1/4xbc$
(c) $3xbc$ (d) $4xbc$

Qu. 4 Simplify $(3x^2y^2)(2x^4y^5)$
Ans (a) $6x^6y^7$ (b) $6x^8y^{10}$ (c) $5x^2y$
(d) $6x^6y^6$

Qu. 5 Simplify $3x(x - 2) - 2x(3x + 4) - 3(x - 2)$
Ans (a) $3x^2 + 17x - 6$ (b) $-3x^2 - 11x - 6$
(c) $-3x^2 - 17x + 6$
(d) $-3x^2 - 15x + 6$

Qu. 6 Evaluate $100(1 + R/100)^2$ when $R = 10$
Ans (a) 121 (b) 1100 (c) 81 (d) 102

Qu. 7 Simplify $5[2(x-1) + 3\{2(x+1) - (x-1)\}]$
Ans (a) $5x + 7$ (b) $25x + 55$
(c) $5x - 15$ (d) $25x + 35$

Qu. 8 Evaluate $R = \dfrac{R_1R_2}{R_1 + R_2}$ when $R_1 = 20$, $R_2 = 30$
Ans (a) 60 (b) 12 (c) 1 (d) 25

Problems 6

1 Given $x = 3$, $y = -2$ evaluate
(a) $3x - 2y$ (b) $5xy$ (c) $3(x + y)$

2 Evaluate the following expressions when $x = 4$
(a) $x^{\frac{1}{2}}$ (b) $x^{-\frac{1}{2}}$ (c) $3(x + 4)^2$

3 Simplify
(a) $7x^2 - 5x + 3x^2 - 10x^2 + 5x + 1$
(b) $(2x^2 + 3xy + 4) + (x^2 + 2xy + 7)$
(c) $(3x^3 + 4x + 1) + (x^3 + 3x^2 + 3x + 1) + (x^3 + 1)$

4 Simplify
(a) $(3x^2 + 7x - 2) - (4 + 3x - 5x^2)$
(b) $(10x^3 + 5x^2 + 3x - 1) - (9x^3 - 4x^2 - 3x + 1)$

5 Simplify
(a) $6x \times -3y \times 4x \times -2y$ (b) $(-3xy)^3$
(c) $\dfrac{30x^3y^4}{6x^2y^2}$ (d) $\dfrac{7x}{20y} \times \dfrac{5x^2}{14y}$

6 Simplify the expressions:
(a) $3x(x+2)-2x(4x+1)-3(1-x)+5x^2$
(b) $4\{3x^2 - x(x+3) - (x^2 + 2x + 3)\}$
(c) $5[3(a-b)-2\{a+b(a-b)+2(b-a)\}$
$+2b(a-b)]$

7 Multiplication and factorization of algebraic expressions

General learning objectives: to multiply and to factorize algebraic expressions involving brackets.

7.1 Multiplication of algebraic expressions

In the last chapter, the introductory chapter on algebra, we applied the laws of algebra, precedence and indices to multiply out and simplify algebraic expressions.

Let us first recapitulate on some important points that should be remembered when multiplying:

1 Rule of signs

$+ \times + = +$ e.g. $+x \times +y = +xy$
$- \times + = -$ e.g. $-x \times +y = -xy$
$+ \times - = -$ e.g. $+x \times -y = -xy$
$- \times - = +$ e.g. $-x \times -y = +xy$

2 Law of indices
Add indices of like terms when multiplying, e.g. $x^2 \times x^5 = x^{2+5} = x^7$

3 When multiplying an expression in brackets, all the individual terms within the brackets must be multiplied,

e.g. $6(2x+1) = 6 \times 2x + 6 \times 1 = 12x + 6$
$-5(x-2) = -5x - 5 \times -2 = -5x + 10$
$3x(x^2 + 2x - 7) = 3x^3 + 6x^2 - 21x$

When we have two or more terms in brackets, e.g. $(x+3)(x+5)$, we must take into account that the terms in the second bracket expression are systematically multiplied by each term within the first bracket expression. For example

$$(x+3)(x+5) = x^2 + 5x + 3x + 15$$
$$= x^2 + 8x + 15$$
$$(2a-3b)(5a+7b) = 10a^2 + 14ab - 15ab - 21b^2$$
$$= 10a^2 - ab - 21b^2$$

Alternatively we can employ a method similar to the long multiplication method used in arithmetic:

$(3x+2)(5x-3) =$

$3x + 2$	
$5x - 3$	
$-9x - 6$	←top expression $\times -3$
$15x^2 + 10x$	←top expression $\times 5x$
$15x^2 + x - 6$	←addition

4 A useful rule for squaring the sum of two terms:

$$(7x-9)^2 = (7x)^2 + 2(7x \times -9) + (-9)^2$$

(square 1st term; add twice the product of the two terms; add square of 2nd term)

$$= 49x^2 - 126x + 81$$

Check by 'long multiplication':

$$\begin{array}{r} 7x - 9 \\ 7x - 9 \\ \hline -63x + 81 \\ 49x^2 - 63x \quad\quad \\ \hline 49x^2 - 126x + 81 \end{array}$$

Examples

1 Multiply $2x^2 - 5x + 3$ by $2x + 3$

Solution

$$(2x+3)(2x^2 - 5x + 3)$$
$$= 4x^3 - 10x^2 + 6x + 6x^2 - 15x + 9$$
$$= 4x^3 - 4x^2 - 9x + 9 \quad \textit{Ans}$$

or using the 'long multiplication' method:

$$\begin{array}{r} 2x^2 - 5x + 3 \\ 2x + 3 \\ \hline 6x^2 - 15x + 9 \\ 4x^3 - 10x^2 + 6x \quad\quad \\ \hline 4x^3 - 4x^2 - 9x + 9 \end{array}$$

2 Multiply $(5x^2 + 3xy + 4y^2)$ by $(2x^2 - xy + y^2)$

Solution

Using the 'long multiplication' method:

$$\begin{array}{rrrrr} & & 5x^2 & +3xy & +4y^2 \\ & & 2x^2 & -\ xy & +\ y^2 \\ \hline & & 5x^2y^2 & +3xy^3 & +4y^4 \\ & -5x^3y & -3x^2y^2 & -4xy^3 & \\ 10x^4 & +6x^3y & +8x^2y^2 & & \\ \hline 10x^4 & +\ x^3y & +10x^2y^2 & -\ xy^3 & +4y^4 \end{array} \quad Ans$$

7.2 Factors of an algebraic expression

A factor of an algebraic expression is a term which will divide into the expression without leaving a remainder.

Examples

1 k is a common factor of the expression $kx + ky$, since both kx and ky contain k. Thus we may write

$$kx + ky = k(x + y)$$

2 5 is a common factor of $5x^2 + 15x + 30$. We can divide the expression throughout by 5 without leaving a remainder, i.e.

$$5x^2 + 15x + 30 = 5(x^2 + 3x + 6)$$

3 $4y$ is a common factor of $4y^3 + 16y^2 + 8y$, i.e.

$$4y^3 + 16y^2 + 8y = 4y(y^2 + 8y + 2)$$

The term **common factor** is used to denote a factor common to all the individual terms in an algebraic expression.

4 Show that $3abc$ is a common factor of

$$9a^2b^2c^2 + 27a^3b^2c + 3abc^2$$

Proof: $3abc$ divides into each of the individual terms, i.e.

$$9a^2b^2c^2 + 27a^3bc + 3abc^2$$
$$= 3abc(3abc + 9a^2 + c)$$

5 Show that $(x + 2)$, $(x - 1)$, $(x - 5)$ are the factors of $x^3 - 4x^2 - 7x + 10$.

Proof: we can show this by multiplying the three factors together; if they are factors the result must be identical to the algebraic expression.

$$(x + 2)(x - 1)(x - 5)$$
$$= (x^2 - x + 2x - 2)(x - 5)$$
$$= (x^2 + x - 2)(x - 5)$$
$$= x^3 + x^2 - 2x - 5x^2 - 5x + 10$$
$$= x^3 - 4x^2 - 7x + 10$$

which is identical to the given algebraic expression. Hence $(x + 2)$, $(x - 1)$, $(x - 5)$ are the factors.

7.3 Factorizing algebraic expressions

In this section we consider some basic methods of finding the factors of algebraic expressions.

7.3.1 Extraction of a common factor

The first step in factorization is to search for the 'highest' common factor contained in the expression.

Examples

1 Extract the highest common factor from the expressions

(a) $12x^2 - 18$ (b) $15y^2 + 20y$
(c) $9a^2 - 36a + 27$

Solution

(a) Both $12x^2$ and -18 contain 6 as a factor (they also contain 2 and 3, but 6 is the highest common factor), so

$$12x^2 - 18 = 6(2x^2 - 3)$$

(b) $5y$ is the only common factor, hence

$$15y^2 + 20y = 5y(3y + 4)$$

(c) 9 is the highest common factor, so

$$9a^2 - 36a + 27 = 9(a^2 - 4a + 3)$$

2 Factorize the following expressions by finding the highest common factor. If the expression has no common factor state so.
(a) $4x^3 + 24x^2 + 16x - 32$
(b) $15x^2 - 21x + 7$
(c) $20a^4b^4 + 30a^3b^3 + 15a^2b^2 + 35ab$

Answers
(a) $4(x^3 + 6x^2 + 4x - 8)$
(b) No common factor
(c) $5ab(4a^3b^3 + 6a^2b^2 + 3ab + 7)$

7.3.2 Factorizing by grouping

When an algebraic expression can be arranged in groups where the individual groups have a compound common factor then this factor is a common factor of the complete expression and the expression may be factorized.

For example, in factorizing expressions containing four terms the method of grouping may often be used. After checking for common factors, arrange the expression into pairs of terms, each pair with its own common factor. Extract the common factor from each pair. A further factor common to each of the groups may then appear. If so we can factorise the expression.

Examples

1 Factorize the expression $ax + bx + ay + by$

Solution

Firstly we can see there is no common factor for all four terms. However, the first two terms have x as a common factor and the last two have y. Group these two pairs in brackets and extract the common factor, i.e.

$$\begin{aligned} ax + bx + ay + by &= (ax + bx) + (ay + by) \\ &= x(a + b) + y(a + b) \end{aligned}$$

Note now that the common factor of $(a + b)$ occurs in both groups. Extracting $(a + b)$ we have,

$$x(a + b) + y(a + b) = (a + b)(x + y) \quad \textit{Ans}$$

2 Factorize $24x - 4y - 12xy + 8$

Solution

First extract the common factor 4:

$$24x - 4y - 12xy + 8 = 4(6x - y - 3xy + 2)$$

The terms $6x$ and $-3xy$ within the brackets have $3x$ as a common factor, so extracting the $3x$, we obtain

$$\begin{aligned} 4(6x - y - 3xy + 2) &= 4\{3x(2 - y) + 2 - y\} \\ &= 4\{3x(2 - y) + (2 - y)\} \\ &= 4(2 - y)(3x + 1) \quad \textit{Ans} \end{aligned}$$

3 Factorize $16ab + 24cd + 32cb + 12ad$

Solution

$$\begin{aligned} &16ab + 24cd + 32cb + 12ad \\ &= 4(4ab + 6cd + 8cb + 3ad) \end{aligned}$$

Common factor of $4b$ Common factor of $3d$

$$\begin{aligned} &= 4\{4b(a + 2c) + 3d(a + 2c)\} \\ &= 4(a + 2c)(4b + 3d) \quad \textit{Ans} \end{aligned}$$

7.3.3 Factorization of binomials: the difference of two squares

A binomial is an algebraic expression containing two terms, e.g. $w + z$, $1 + x^2$, $9x^2 - 25y^2$ are binomials.

The majority of binomials, apart from a common factor, do not factorize any further. However, there is a special and important case when the binomial consists of the 'difference of two squares', that is the binomial is of the form:

$$X^2 - Y^2$$

For this case the binomial has the factors $(X + Y)$ and $(X - Y)$:

$$X^2 - Y^2 = (X + Y)(X - Y)$$

which we can easily check:

$$\begin{aligned} (X + Y)(X - Y) &= X^2 - XY + YX - Y^2 \\ &= X^2 - Y^2 \end{aligned}$$

Examples

1 Factorize the following. If the binomial cannot be factorized state so.
(a) $4x^2 - 9$ (b) $64x^2 - 144y^2$
(c) $49z^2 + 25b^2$

Solution

(a) $4x^2$ is a perfect square: $4x^2 = 2x \times 2x$
9 is a perfect square: $9 = 3 \times 3$
so $4x^2 - 9 = (2x + 3)(2x - 3)$ *Ans*

(b) Both terms are perfect squares, but first extract the highest common factor:

$$64x^2 - 144y^2 = 16(4x^2 - 9y^2)$$
$$= 16(2x + 3y)(2x - 3y) \quad Ans$$

(c) This binomial cannot be factorized. Although both terms are perfect squares the expression is a *sum* not a difference.

2 Using the result $x^2 - y^2 = (x + y)(x - y)$, evaluate
(a) $1000^2 - 999^2$ (b) $0.501^2 - 0.499^2$

Answers

(a) $1000^2 - 999^2 = (1000 + 999)(1000 - 999)$
$= 1999 \times 1 = 1999$

(b) $0.501^2 - 0.499^2$
$= (0.501 + 0.499)(0.501 - 0.499)$
$= 1 \times 0.002 = 0.002$

Test and problems 7

Multiple choice test: MT7

Answer block:

Question No.	0	1	2	3	4	5	6	7	8
Answer	c								

Enter your answer, that is a, b, c or d in the column under the question number in the answer block above. Note that question Qu 0 has already been worked out and the answer inserted.

Qu. 0 Factorize the expression $21x^2 + 49x^2y^2$
Ans (a) $x^2(21 + 49y^2)$ (b) no factors
(c) $7x^2(3 + 7y^2)$ (d) $7(x^2 + 7x^2y^2)$

Solution

$21x^2 + 49x^2y^2$ has $7x^2$ as a common factor, so
$21x^2 + 49x^2y^2 = 7x^2(3 + 7y^2)$ *Ans*

so the correct answer is (c) and we insert 'c' under Qu. 0 in the answer block as shown.

Now carry on with the test.

Qu. 1 Multiply $(1 - x)$ by $(2 - 3x)$
Ans (a) $2 - 5x - 3x^2$ (b) $3x^2 - 5x - 2$
(c) $3x^2 - 5x + 2$ (d) $2 + 5x - 3x^2$

Qu. 2 Multiply $x^2 + 3x + 4$ by $5x + 1$
Ans (a) $5x^3 + 16x^2 + 23x + 4$
(b) $5x^2 + 16x + 27$
(c) $5x^3 + x^2 + 17x + 4$
(d) $5x^3 + 15x^2 + 20x + 4$

Qu. 3 Simplify $(x + 1)(x + 2) - (x - 1)(x - 2)$
Ans (a) 0 (b) $6x$
(c) $6x + 4$ (d) $x^2 - 3x + 4$

Qu. 4 Factorize $25x^2 + 125$
Ans (a) $5(5x^2 + 25)$ (b) $25(x^2 + 5)$
(c) $5(x + 5)(x - 5)$ (d) $5(x^2 + 25)$

Qu. 5 Factorize $2a^2 + 3ab + 2ac + 3bc$
Ans (a) $(a + b)(2a + 3c)$ (b) no factors
(c) $(3a + 2b)(a + c)$ (d) $(2a + 3b)(a + c)$

Qu. 6 State whether or not 6, $(y - 7)$, $(x + y)$ are factors of the expression $6y^2 - 42y + 6xy - 42x$
Ans (a) only 6 (b) all three are factors
(c) $(x + y)$ is not a factor
(d) $(y - 7)$ is not a factor

Qu. 7 Factorize $x^2 - y^2 + x + y$
Ans (a) no factors (b) $(x - y)(x + y)$
(c) $(x - y)(x + y + 1)$
(d) $(x + y)(x - y + 1)$

Qu. 8 Simplify $(2x + 1)^2 - (x + 5)^2$
Ans (a) $3(x + 2)(x - 4)$ (b) $3x^2 - 6x + 24$
(c) $2x^2 + 6x + 24$ (d) $(x - 4)(3x + 5)$

Problems 7

1 Multiply out
(a) $(2x + 5)^2$ (b) $(x + 3)(x - 7)$
(c) $(5a - b)(2a - b)$

2 Simplify the expressions
(a) $3x(x - 2) + 5x(x + 2)$
(b) $(x + y)(2x^2 + 3x + 4xy + y + y^2)$
$-(3x^2 + 6x^2y + y^2 + y^3)$

3 Show that $(x + 1)(x + 2)(x + 3)$ are the factors of the expression $x^3 + 6x^2 + 11x + 6$

4 Factorize
(a) $28x^2 + 7x$ (b) $27x - 54y$ (c) $36a^2b^2 + 9ab$

5 Factorize
(a) $1 - y^2$ (b) $49x^2 - 25$ (c) $100x^2 - 25y^2$

6 Obtain the factors of the expressions
(a) $3p + ay + 3y + ap$
(b) $24t - 12st - 4s + 8$
(c) $x^3 - qx^2 - 4q^2x + 4q^3$

7 (a) Show that $x^3 + y^3 = (x + y)(x^2 - xy + y^2)$
and hence factorize $27a^3 + 125b^3$
(b) Factorize $12x - 2y - 6xy + 4$

8 The solution of simple and simultaneous equations

General learning objectives: to solve algebraically simple linear and simultaneous equations and to relate to the solution of practical problems.

8.1 Equations, identities and inequalities

An equation is a mathematical statement expressing an equality. An equation always contains an equals sign '=' and equates an algebraic expression to another algebraic expression.

In this chapter we consider the solution and the formation of two important types of equations: simple linear equations and linear simultaneous equations containing two unknowns.

8.1.1 Simple linear equations

A simple linear equation contains only one unknown (hence the qualification 'simple') and no other power of the unknown, i.e. no x^2, x^3, etc. terms. The qualification 'linear' is for the latter reason; if we were to solve a simple linear equations by graphical methods the plot of the equation would be 'linear': a straight line.

The following are examples of simple linear equations:

$$5x + 3 = 23 \quad (1)$$
$$2y + 7 = y + 10 \quad (2)$$
$$\tfrac{1}{3}(z - 4) = \tfrac{1}{5}(z + 6) \quad (3)$$

In equation (1), x is used to denote the unknown, in equation (2) y is used and in equation (3) z.

The solutions of these equations, that is the value of the unknown which satisfies the equality condition, are:

for equation (1): $x = 4$
for equation (2): $y = 3$
for equation (3): $z = 19$

We may check that each value 'satisfies' its respective equation by substituting the solution back into the equation. If the solution is the correct one, then both sides of the equation must be equal, i.e. left-hand side (LHS) of equation = right hand side (RHS) of equation.

so for equation (1): LHS $= (5 \times 4) + 3 = 23$
RHS $= 23$
so equation balances and hence $x = 4$ is the solution;

for equation (2): LHS $= (2 \times 3) + 7 = 13$
RHS $= 3 + 10 = 13$
hence LHS = RHS and $y = 3$ is the solution;

for equation (3): LHS $= \tfrac{1}{3}(19 - 4) = \tfrac{1}{3} \times 15 = 5$
RHS $= \tfrac{1}{5}(19 + 6) = \tfrac{1}{5} \times 25 = 5$
hence LHS = RHS and $z = 19$ is the solution

8.1.2 Linear simultaneous equations

Simultaneous linear equations contain two or more unknowns, but no powers of the unknowns. The equations

$$3x + 5y = 21 \quad (4a)$$
$$2x - 3y = -5 \quad (4b)$$

are an example of a pair of simultaneous equations relating the two unknowns, x and y. If we have two unknowns we must always have two independent equations to obtain solutions for the unknowns.

The solutions for equations (4a) and (4b) are

$$x = 2, \qquad y = 3$$

We can check by substituting these values into the two equations:

LHS of (4a): $(3 \times 2) + (5 \times 3) = 6 + 15 = 21$ = RHS of (4a)

LHS of (4b): $(2 \times 2) - (3 \times 3) = 4 - 9 = -5$ = RHS of (4b) we see that both equations balance so $x = 2$ and $y = 3$ are the solutions.

In general we require as many independent equations as there are unknowns. If we had fewer we could not find a unique solution for each unknown; there would be an infinite number. Thus, for example, if we were solving for three unknowns we would require three independent simultaneous equations. Check by substituting that the solutions for the three simultaneous equations:

$$x - y + 7z = 36$$
$$4x + 9y - 3z = 15$$
$$3x + 2y + z = 18$$

are given by $x = 3$, $y = 2$, $z = 5$

Table 8.1 lists the meaning of a number of mathematical symbols used to express various equality and inequality relations in concise form.

When an equality expression holds for all values of the individual terms, the 'equation' is known as an **identity** and the equals sign is replaced by the identity symbol $\equiv$.
For example, the following are identities:

$$(x + y)^2 \equiv x^2 + 2xy + y^2$$
$$(x^2 - y^2) \equiv (x + y)(x - y)$$
$$(x^3 + y^3) \equiv (x + y)(x^2 - xy + y^2)$$

since LHS = RHS for all values of x and y. An identity, unlike a normal equation, holds for all numerical values given to the 'unknowns'. For example, the equation

$$9x + 7 = 4x + 37$$

Table 8.1 Equality and inequality symbols

Symbol	Meaning
$=$	is equal to
$\equiv$	is identical to
$\approx$	is approximately equal to
$\propto$	is proportional to
$>$	is greater than
$\geqslant$	is greater than or equal to
$<$	is less than
$\leqslant$	is less than or equal to
$\gg$	is much greater than
$\ll$	is much less than
$\neq$	is not equal to

has a single (unique) solution, $x = 6$. Only for this value can the equation be satisfied, so LHS = RHS.

To write in mathematical terms when an expression is greater than, less than, etc. another expression, we employ the inequality symbols summarized in Table 8.1. For example

$y > x + 2$ means that y *is greater than* $(x + 2)$.
$x + 6 < 0$ means that $(x + 6)$ *is less than* 0, so in this case x must always be less (more negative) than -6.
$y \gg x$ means that y *is much greater than* x; in practice this means y is at least an order of magnitude greater than x, e.g. at least 10 times.
$y \leqslant 16$ means that y *is less than or equal to* 16, so in this case y can have values of 16 or less.

8.2 The solution of linear equations with one unknown: simple equations

The general rule when manipulating an equation is to preserve the equality. If an operation is made on the LHS of an equation the identical operation must be carried out on the RHS.

To solve simple equations we manipulate both sides of the equation so as to gather terms in the unknown on one side and numbers on the other: we transfer all x (unknown) terms to, say, the LHS and all number terms to the RHS; we then divide the algebraic sum of the numbers, the total sum, on the RHS by the total number of xs we have on the LHS to finally determine the value of x.

The following two rules form the basis for solving simple equations:

1 *The same quantity may be added to or subtracted from an equation, since equality will be maintained.*
For example, solve $3x - 5 = 2x + 9$

Solution

First add 5 to both sides:

$$3x - 5 + 5 = 2x + 9 + 5$$

so $$3x = 2x + 14$$

Next subtract $2x$ from both sides:

$$3x - 2x = 2x + 14 - 2x$$

so $\quad x = 14$ giving us the solution

An equivalent way of effecting the addition/subtraction to both sides of an equation is:

If we transfer a quantity from one side of an equation to the other side, then we must change the sign of the quantity transferred. So illustrating this technique, let us solve

$$7x + 10 = 6x - 33$$

Solution

$$7x + 10 = 6x - 33 - 10$$

i.e. $\quad 7x = 6x - 33 - 10 = 6x - 43$

on transferring $+10$ from LHS to RHS.
Now transferring $6x$ from the RHS to LHS we have

$$-6x + 7x = 6x - 43$$

so $\quad x = -43$, our solution.

2 *Both sides of an equation may be multiplied by or divided by any number other than zero*

For example, solve $\frac{1}{6}(5x + 3) = \frac{1}{2}(3x - 2) + \frac{1}{3}x$

Solution

First multiply both sides of the equation by 6 to remove fractions:

$$6 \times \frac{1}{6}(5x + 3) = 6 \times \frac{1}{2}(3x - 2) + 6 \times \frac{1}{3}x$$

so $\quad (5x + 3) = 3(3x - 2) + 2x$

$$5x + 3 = 9x - 6 + 2x = 11x - 6$$

Next transfer x terms to one side (say the LHS):

$$5x - 11x + 3 = -6$$

and then the number terms to the other, RHS in this case:

$$5x - 11x = -6 - 3$$
$$-6x = -9$$

Finally divide both sides by -6 (the number of x's),

$$x = \frac{-9}{-6} = \frac{9}{6} = 1\tfrac{1}{2} \quad \textit{Ans}$$

Examples

1 Solve the simple equation: $8x - 9 = 3x + 11$

Solution

Add 9 to both sides:

$$8x - 9 + 9 = 3x + 11 + 9$$
$$8x = 3x + 20$$

Subtract $3x$ from both sides:

$$8x - 3x = 3x + 20 - 3x$$
$$5x = 20$$

Divide both sides by 5:

$$\frac{5x}{5} = \frac{20}{5}$$
$$x = 4 \quad \textit{Ans}$$

2 Solve $\frac{1}{4}(8x + 1) = \frac{3}{8}(5x - 1)$

Solution

Multiply both sides by 8 to remove fractions:

$$8 \times \frac{1}{4}(8x + 1) = 8 \times \frac{3}{8}(5x - 1)$$
$$2(8x + 1) = 3(5x - 1)$$
$$16x + 2 = 15x - 3$$

Subtract $15x$ from both sides:

$$16x + 2 - 15x = 15x - 3 - 15x$$
$$x + 2 = -3$$

Subtract 2 from both sides:

$$x + 2 - 2 = -3 - 2$$

so finally $\quad x = -5 \quad$ *Ans*

3 Solve $2\{x + 3(x + 2) + 5\} = \frac{1}{3}\{2x - 5(x - 1)\}$

Solution

First multiply both sides of the equation by 3 to remove fractions, then simplify by clearing the brackets:

$$3 \times 2\{x + 3(x + 2) + 5\} = 3 \times \frac{1}{3}\{2x - 5(x - 1)\}$$

$$6\{x + 3x + 6 + 5\} = \{2x - 5x + 5\}$$
$$6(4x + 11) = -3x + 5$$
$$24x + 66 = -3x + 5$$

and on transferring x terms to LHS, number to RHS, we obtain:

$$24x + 3x = 5 - 66$$
$$27x = -61$$
$$x = -\frac{61}{27} = -2\frac{7}{27} \quad Ans$$

4 Show that the equation:

$$\frac{(x-3)(x+4)}{5} = \frac{(2x-1)(x+3)}{10}$$

reduces to a simple equation and hence solve for x

Solution

On multiplying both sides by 10 to remove fractions we obtain

$$2(x-3)(x+4) = (2x-1)(x+3)$$

and multiplying out the bracket terms on both sides, we have

$$2(x^2 + 4x - 3x - 12) = (2x^2 + 6x - x - 3)$$
$$2x^2 + 2x - 24 = 2x^2 + 5x - 3$$

If we now subtract $2x^2$ from both sides, we obtain the simple equation:

$$2x - 24 = 5x - 3$$

and solving this:

$$2x - 5x = -3 + 24$$
$$-3x = 21$$
$$x = \frac{21}{-3} = -7 \quad Ans$$

5 Solve the equation

$$\frac{9}{5x-6} = \frac{4}{x+1}$$

Solution

Here we have an example of an algebraic fraction type equation with algebraic expression in the denominator. We can solve the equation by first converting it into a more convenient form Invert both LHS and RHS. This is equivalent to dividing both sides into 1 and as we have performed the same operation on both sides we have preserved the equality.
So

$$\frac{5x-6}{9} = \frac{x+1}{4}$$

now multiply both sides by 36 to remove the number fractions,

$$\frac{36(5x-6)}{9} = \frac{36(x+1)}{4}$$

hence

$$4(5x-6) = 9(x+1)$$
$$20x - 24 = 9x + 9$$
$$20x - 9x = 9 + 24$$
$$11x = 33$$
$$x = 3 \quad Ans$$

Note: technique of cross-multiplication
An extremely useful result that can be employed when equations are expressed in fractional or ratio form (such as those of examples 4 and 5) is to equate the *cross-product terms*, i.e. if

$$\frac{A}{B} = \frac{C}{D} \qquad \text{then } AD = BC$$

AD and BC are the cross-products formed by the numerator of one side of the equation with the denominator of the other side. The above result can easily be proved by multiplying both sides of the 'ratio' equation by BD:

$$\frac{A}{\not B} \times \not B D = \frac{C}{\not D} \times B\not D \quad \text{so} \quad AD = BC$$

Thus for example, the equation:

$$\frac{9}{(5x-6)} = \frac{4}{(x+1)}$$

can be immediately converted to: $9(x+1) = 4(5x-6)$

8.3 Simple equation formation and solution for some practical problems

In the solution of many practical problems our first step is to form an equation or a number of equations from the given information, using letter symbols to denote the unknown quantities we want to determine. In science, for example, we apply our knowledge of experimentally determined or theoretical laws to form equations linking the unknowns with the numerical data supplied. We form equations and then proceed to solve them.

Below we consider some practical problems involving one unknown and which can be solved by setting up a simple equation.

Examples

1 Figure 8.1 shows a simple circuit. Our problem is to find the current flowing in the circuit when the switch S is closed.

Solution

Denote the unknown electrical current by I. To find I we must establish an equation linking I with the given data:

Battery voltage $V_b = 9$ volts
Resistor values: $R_1 = 15$ ohms,
$R_2 = 35$ ohms

We use two laws to construct the equation:

Ohm's law: voltage drop across a resistor equals the product of current flowing through the resistor and its resistance, i.e. $V = RI$

and the fact that in a series circuit,

applied voltage = sum of voltage drops across individual resistors

Thus applying Ohm's law, we obtain

voltage drop across 15 ohm resistor $= R_1 I = 15I$
voltage drop across 35 ohm resistor $= R_2 I = 35I$

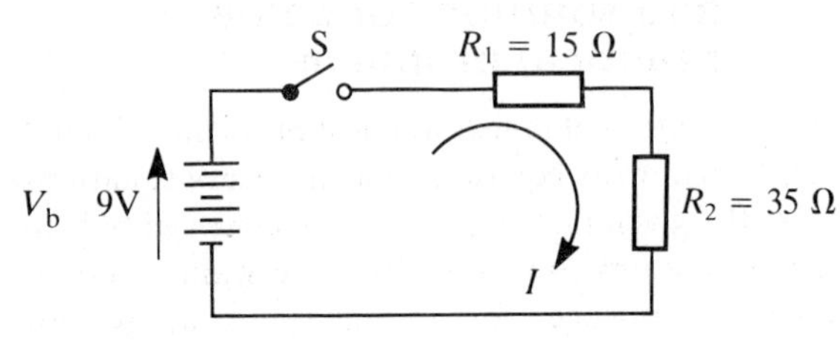

Figure 8.1.

and also since the applied voltage is the battery voltage of 9 volts,

$$9 = 15I + 35I$$

so

$$9 = 50I, \quad I = \frac{9}{50} = 0.18 \text{ amperes} \quad \textit{Ans}$$

2 The weekly wage bill for a workforce of 200 is £33,300. The workforce is composed of skilled and semi-skilled workers. Skilled workers each earn £205 per week and semi-skilled £150. Determine the number of skilled workers employed.

Solution

Let x = number of skilled workers, then $200 - x$ = number of semi-skilled workers

The weekly wage bill for,

$$\text{skilled workers} = 205 \times x$$
$$\text{semi-skilled} = 150 \times (200 - x)$$

and equating the sum of these two to the total of 33 300, we obtain the equation:

$$205x + 150(200 - x) = 33\,300$$
$$205x + 30\,000 - 150x = 33\,300$$
$$205x - 150x = 33\,300 - 30\,000$$
$$55x = 3300$$
$$x = \frac{3300}{55} = 60 \quad \textit{Ans}$$

3 Using the compound interest formula,

$$A = P(1 + R/100)^T$$

where $A = P + I$, P = sum invested, I = interest
R = rate of interest in % per annum
T = time in years of investment

Determine (a) the rate R which would enable the sum of £1000 to at least double in 8 years;
(b) the time taken for £500 to grow to £1500 if invested at a rate of interest of 7% per annum.

Solution

(a) Since £ 1000 with interest is to at least double, we have $A = £2000$. Substituting therefore

$A = 2000$, $P = 1000$ and $T = 8$ in the formula, we obtain

$$2000 = 1000\,(1 + R/100)^8$$

so on dividing both sides by 1000 and then taking the 8th root, we have,

$$\sqrt[8]{2} = (1 + R/100)$$

On evaluating $\sqrt[8]{2}$ using a calculator or log tables,

$$\sqrt[8]{2} = 1.09051 = (1 + R/100)$$
$$1.09051 - 1 = R/100$$
$$R = 100 \times 0.09051 = 9.051\% \quad \textit{Ans}$$

(b) In this case $A = 1500$, $P = 500$, $R = 7$ and on substituting into the formula, we obtain

$$1500 = 500(1 + 7/100)^T$$
$$3 = (1.07)^T \quad \text{on dividing both sides by 500.}$$

The easiest way to solve this equation is to take logarithms of both sides,

$$\log 3 = \log 1.07^T = T \log 1.07$$

hence we can now find T directly,

$$T = \frac{\log 3}{\log 1.07} = \frac{0.4771}{0.02938} = 16.24 \text{ years} \quad \textit{Ans}$$

4 An airliner flies at a constant airspeed for 3 hours against a head on wind of 40 km/h (this means the wind reduces the effective speed of the plane over land by 40 km/h). The distance covered in this time is 2100 km. Determine the airspeed of the plane.

Solution

Let the 'plane's airspeed $= v$ km/h, then the actual speed of the 'plane over the ground is $(v - 40)$ km/h and on using the general result,

$$\text{distance} = \text{speed} \times \text{time}$$

we have $2100 = (v - 40) \times 3$

so dividing both sides by 3

$$\frac{2100}{3} = v - 40$$
$$700 = v - 40$$
$$700 + 40 = v$$
$$v = 740 \text{ km/h} \quad \textit{Ans}$$

8.4 The solution of simultaneous linear equations in two unknowns

Simultaneous as distinct from simple equations contain two or more unknowns. The equations are termed 'simultaneous' because the solutions must simultaneously satisfy the set of equations. We previously stated that in order to solve simultaneous equations and obtain a unique solution the number of equations must equal the number of unknowns. Hence if we have two unknowns, x and y say, then we require two independent equations relating x and y and we solve these 'simultaneously' to find x and y.

In this section we consider two basic methods for solving simultaneous equations in two unknowns.

8.4.1 Method of elimination

In this method one of the unknown terms, either the x or y term, is 'eliminated' by multiplying the two equations in turn by the appropriate numbers to make the x or y term have the same coefficient in both equations. We can then subtract one equation from the other to effect the elimination of one unknown and which results in leaving a simple equation. The simple equation is then solved.

It is much easier to see the idea of the method by example:

Solve

$$5x + 4y = 7 \qquad (1a)$$
$$3x - 7y = 23 \qquad (1b)$$

To eliminate x term from those equations,

multiply (1a) throughout by 3: $15x + 12y = 21$
multiply (1b) throughout by 5: $15x - 35y = 115$

and now subtract $47y = -94$

so $$y = \frac{-94}{47} = -2$$

To find the other unknown x, substitute $y = -2$ in either equation (1a) or (1b). If we substitute into (1a), we obtain

$$5x + 4(-2) = 7$$
$$5x - 8 = 7$$
$$5x = 7 + 8 = 15$$

so $x = \dfrac{15}{5} = 3$

Hence our solutions are $x = 3, y = -2$

8.4.2 Method of substitution

In this method one equation is used to express one unknown in terms of the other, then this equation is substituted for the appropriate unknown in the second equation. The resulting equation after this substitution is reduced to a simple equation, which is then solved. For example,

$$2x - 3y = 4 \qquad (2a)$$
$$6x + 5y = 40 \qquad (2b)$$

Using equation (2a) to express x in terms of y, we have

$$2x = 4 + 3y$$
$$x = \frac{1}{2}(4 + 3y) \qquad (3)$$

and on substituting for x in equation (2b), we obtain

$$6 \times \frac{1}{2}(4 + 3y) + 5y = 40$$
$$3(4 + 3y) + 5y = 40$$
$$12 + 9y + 5y = 40$$

so $14y = 40 - 12 = 28$

$$y = \frac{28}{14} = 2$$

Finally now substitute $y = 2$ in equation (3) to find x,

$$x = \frac{1}{2}(4 + 3 \times 2) = \frac{1}{2}(4 + 6) = 5$$

so the solutions are $x = 5, y = 2$ *Ans*

Examples

1 Solve the simultaneous equations:

$$7x + 4y = 25 \qquad (1)$$
$$4x + 3y = 10 \qquad (2)$$

Solution

Using the elimination method to eliminate y from the equations:

multiply (1) by 3: $21x + 12y = 75$
multiply (2) by 4: $16x + 12y = 40$

and subtracting: $5x \qquad = 35$

so $x = \dfrac{35}{5} = 7$ and on substituting $x = 7$ in (1),

$$49 + 4y = 25$$
$$4y = 25 - 49 = -24$$
$$y = -\frac{24}{4} = -6$$

Thus the solutions are $x = 7, y = -6$ *Ans*

2 Solve the equations:

$$x - y = 2.2 \qquad (3)$$
$$3.6x + 4.2y = 7.9 \qquad (4)$$

Solution

Using (3) to express x in terms of y, we have

$$x = 2.2 + y \qquad (5)$$

and on substituting this value for x in equation (4),

$$3.6(2.2 + y) + 4.2y = 7.9$$
$$(3.6 \times 2.2) + 3.6y + 4.2y = 7.9$$
$$7.8y = 7.9 - (3.6 \times 2.2)$$
$$= 7.9 - 7.92 = 0.02$$

so $y = \dfrac{0.02}{7.8} = 0.002\,56$

and using (5), $x = 2.2 + 0.002\,56 = 2.202\,56$
Hence the solutions are: $x = 2.202\,56$,
$y = 0.002\,56$ *Ans*

8.5 Simultaneous equation formation and solution for some practical problems

In section 8.3 we considered setting up equations for problems involving one unknown. Many practical problems can be defined and solved only by involving two or more unknowns. In this section we construct equations – simultaneous equations – for problems which can be defined by considering two unknowns.

Our task is to set up two independent simultaneous equations linking the two unknowns using the facts and data supplied. Each of these two equations must be independent in the sense that it takes into account 'different' parts of the supplied data and the two equations together cover all the relevant information given in the problem. To construct the equations we apply, for example, physical or financial facts and laws, formulae, etc. to link the supplied data with, of course, common-sense reasoning.

Examples

1 A metal alloy consists by mass 3 parts of metal X and 4 parts of metal Y. Metals X and Y cost respectively £5 and £2 per kilogram. Calculate the material cost in producing an alloy bar of mass 10 kg.

Solution

First let us find the respective masses of metal in the 10 kg bar. Let,

x = mass of metal X, in kg
y = mass of metal Y, in kg

then, as the total mass of the bar is 10 kg, we have

$$x + y = 10 \qquad (1)$$

The second equation is obtained using the fact that the alloy is made up of 3 parts of metal X and 4 parts metal Y, so

$$\frac{x}{y} = \frac{3}{4} \quad \text{or} \quad 4x = 3y \qquad (2)$$

From (1) we have, $x = 10 - y$
and on substituting this value in (2), we obtain

$$4(10 - y) = 3y$$
$$40 - 4y = 3y$$

so $$40 = 3y + 4y = 7y$$

$$y = \frac{40}{7} = 5.714$$

and $$x = 10 - y = 10 - 5.714 = 4.286$$

Thus the material cost of the bar with X costing £5 and Y costing £2 per kg, is

$$\text{cost} = 5x + 2y = 5 \times 4.286 + 2 \times 5.714$$
$$= £32.86 \quad \textit{Ans}$$

Note: The problem could be solved without setting up the two simultaneous equations by using the following reasoning:

Since the bar consists 3 parts X, 4 parts Y, divide the 10 kg into $(3 + 4) = 7$ parts, then

metal X = 3 parts

$$\text{so } x = \frac{10}{7} \times 3 = 4\frac{2}{7} \quad \text{or} \quad 4.286\,\text{kg}$$

metal Y = 4 parts

$$\text{so } y = \frac{10}{7} \times 4 = 5\frac{5}{7} \quad \text{or} \quad 5.714\,\text{kg}$$

2 The number of workers employed, the annual income and net profit for a manufacturing company over two successive years are summarized in the table below:

Year	Number of workers	Gross income	Net profit
1	30	£651 000	£66,000
2	35	£998, 400	£209,400

The net profit is calculated after deducting annual overheads, materials, expenses, etc. and also the wages bill from the gross income.

From year 1 to year 2 the average worker's salary increased by 20% and overhead, etc. expenses by 30%.

Determine the average worker's salary and total expenses on overheads, etc. for year 1 and year 2.

Solution

Let the average worker's salary in year 1 be x (in £1000 units) and the total expenses in year 1 be y (again in £1000 units).

Then in year 1:
total salaries + total expenses = gross income less net profit

$$30x + y = 651 - 66 = 585$$

i.e. $$30x + y = 585 \qquad (1)$$

In year 2
due to wage inflation of 20%,

$$\text{worker's average salary} = x \times \frac{100+20}{100} = 1.2x$$

due to overheads, etc. inflation of 30%

$$\text{total expenses} = y \times \frac{100+30}{100} = 1.3y$$

so the 2nd year salary bill $= 1.2x \times 35 = 42x$
and therefore the new equation for
total salaries + total expenses = gross income less net profit

is $42x + 1.3y = 998.4 - 209.4 = 789$ (2)

We now solve equations (1) and (2):

from (1) $y = 585 - 30x$ (3)

and substituting this for y in (2), we obtain

$$42x + 1.3(585 - 30x) = 789$$
$$42x + 760.5 - 39x = 789$$
$$3x = 789 - 760.5 = 28.5$$

so $x = \frac{28.5}{3} = 9.5$ (in £1000 units)

and using (3), $y = 585 - (30 \times 9.5)$
$= 585 - 285 = 300$ (in £1000 units)

Thus for year 1:

average worker's wage,
$x = £9.5 \times 1000 = £9{,}500$
overhead, etc. expenses,
$y = £300 \times 1000 = £300{,}000$

For year 2:

average worker's wage
$= 1.2 \times x = £9{,}500 \times 1.2 = £11{,}400$
overhead, etc. expenses
$= 1.3 \times y = £300{,}000 \times 1.3 = £390{,}000$

Test and problems 8

Multiple choice test: 8

Answer block:

Question No.	0	1	2	3	4	5	6	7	8
Answer	c								

Enter your answer, that is a, b, c or d in the column under the question number in the answer block above. Note that question Qu 0 has already been worked out and the answer inserted.

Qu. 0 The total daily wage bill for a site employing 20 workers is £760. The 20 workers consist of skilled and semi-skilled who earn respectively £40 and £30 per day. Determine the number of skilled workers on the site
Ans (a) 8 (b) 12 (c) 16 (d) 10

Solution

Let x = number of skilled workers, then

$20 - x$ = number of semi-skilled.

The total daily bill is

$$40x + 30(20 - x) = 760$$
$$40x + 600 - 30x = 760$$
$$10x = 760 - 600 = 160$$

so $x = \frac{160}{10} = 16$ *Ans*

so the correct answer is (c) and we insert 'c' in the answer block under Qu 0 as already shown

---

Now carry on with the test.

Qu. 1 Solve the equation, $3(x - 4) + 6 = 24$
Ans (a) 6 (b) 12 (c) 4 (d) 10

Qu. 2 Solve the equation, $\frac{5x+3}{6} = \frac{x+6}{2}$

Ans (a) 6 (b) $7\frac{1}{2}$ (c) 4 (d) -3

Qu. 3 Solve the equations:

$$x - y = 8$$
$$x + y = 0$$

Ans (a) $x = 8$, $y = 0$
(b) $x = 4$, $y = -4$
(c) $x = 10$, $y = 2$
(d) $x = 0$, $y = -8$

Qu. 4 Solve the equations:

$$5x + 4y = 2$$
$$3x + 7y = 15$$

Ans (a) $x = -2$, $y = 3$
(b) $x = 2$, $y = -3$

(c) $x = 3, \quad y = 2$
(d) $x = 4, \quad y = -2$

Qu. 5 The solutions for the simultaneous equations

$$x + y + z = 6$$
$$x - y - z = 0$$
$$x + 3y + 2z = 11$$

are (a) $x = 1, \quad y = 2, \quad z = 3$
(b) $x = 2, \quad y = 1, \quad z = 3$
(c) $x = 3, \quad y = 2, \quad z = 1$
(d) $x = 0, \quad y = 4, \quad z = 2$

Qu. 6 The sum of two numbers is 30 and their difference is 4. Determine the two numbers
(a) 14, 16 (b) 34, -4
(c) 12, 18 (d) 17, 13

Qu. 7 Solve the equation, $\dfrac{x-3}{4} + \dfrac{2x+1}{3} = \dfrac{9x+7}{12}$

Ans (a) 1 (b) 4 (c) 6 (d) -6

Qu. 8 An athlete starts from a given point and runs at an average speed of 5 m/s. 15 minutes later a cyclist leaves the same point and after cycling for 20 minutes overtakes the runner. Calculate the average speed of the cyclist
Ans (a) 8.75 m/s (b) 11.6 m/s
(c) 15 m/s (d) 12 m/s

Problems 8

1 Solve the following simple equations:

(a) $\dfrac{5}{8}x = 10$ (b) $6x + 4 = 28$

(c) $3x - 5 = x + 5$ (d) $\dfrac{1}{4}(2x - 1) = \dfrac{1}{3}(x + 1)$

(e) $\dfrac{1}{x} + \dfrac{1}{60} = \dfrac{1}{20}$

2 Solve the equations

(a) $\dfrac{3}{2x+1} = \dfrac{4}{3x+2}$

(b) $6.6(x - 0.2) + 9.2 - 2.6x = 0$

3 The equation $s = ut + \frac{1}{2}at^2$ is the formula for the distance s metres travelled by a body with an initial velocity of u metres per second (that is the speed of the body at time $= 0$) when accelerated at a rate of a metres per second per second for a time of t seconds.

If it is known that the distance travelled in $t = 10$ seconds with a body of initial velocity $u = 15$ m/s is $s = 250$ m, determine the acceleration a.

4 Solve the equations,

(a) $\dfrac{x}{3} + \dfrac{x-2}{4} = \dfrac{3x+5}{12}$

(b) $\dfrac{x}{6} + \dfrac{x}{7} = 13$

5 It is found experimentally that the length of many metals varies with temperature according to the formula:

$$l_2 = l_1[1 + \alpha(T_2 - T_1)]$$

where l_2 = length at T_2°C, l_1 = length at T_1°C and α = a constant for a given material, known as the coefficient of linear expansion.

In an experiment to determine the α of steel the following results were recorded: length of a steel bar at 20°C = 180 mm; increase in length when the bar was immersed in a steam bath at 100°C = 0.17 mm. Using these results determine α.

6 Solve the simultaneous equations

(a) $5x + 4y = 2$
$3x + 7y = 15$

(b) $6.6x - 0.7y = 8.29$
$5x + 1.2y = 10.26$

7 The velocity of a body experiencing a constant acceleration of a m/s^2 is given by the formula:

$$v = u + at$$

where u = its initial velocity (velocity at time = 0)
t = time in seconds

The following results were recorded for such a body:

when $t = 10$ s, $v = 14$ m/s
$t = 20$ s, $v = 19$ m/s

Determine the initial velocity u and acceleration a for this body.

8 Solve the simultaneous equations

(a) $2x - 3y = -21$
$5x + 7y = 20$

(b) $1.2a + 2.2b = 2.36$
$0.7a - 1.5b = -0.85$

9 The electrical current in an electronic device is given by the formula: $I = a + bV$ amperes, where $V =$ control voltage and a and b are constants. Given $I = 0.25$, when $V = -4$ and $I = 0.050$ when $V = -8V$, calculate the constants a and b.

9 The evaluation and transformation of formulae

General learning objectives: to evaluate and transform formulae.

9.1 The evaluation of formulae by substitution of given data

A mathematical formula is an equation that expresses a given quantity (usually known as the subject of the formula) in terms of the other quantities (often known as variables) to uniquely define the quantity. Formulae may be derived by mathematical reasoning, by applying practical experience, the laws of science, economics, etc.

Once having established a general formula the subject may be evaluated by substituting in the expression the given values for the variables. In evaluating formulae, as in all calculations, the following points should always be bourne in mind:

1. Ensure that the values for the quantities to be substituted into the formula are in the correct units.
2. Make sure you understand the meanings of the variable symbols; do not confuse one symbol with another.
3. Make a rough calculation to check your 'accurate' result, say, obtained with an electronic calculator.
4. Ask yourself 'is the answer physically realistic?' If not, check through the supplied data and rework the calculation. If the result is still unrealistic, reject it, stating your reason as to why.

Examples

1 The volume v and mass m of the solid cylinder shown in Figure 9.1 are given by the formulae

$$v = \pi r^2 h; \quad m = \rho v$$

where r = cylinder radius in metres, m
h = cylinder height in metres
ρ = density of cylinder material in kilograms per cubic metre, kg/m^3

Calculate the volume and mass of a solid brass cylinder of radius 50 mm and height 215 mm. Take $\pi = 3.142$ and the density of brass $\rho = 8.5 \times 10^3$ kg/m^3

Solution

$r = 50\text{ mm} = 50 \times 10^{-3}\text{ m}$, as $1\text{ m} = 1000\text{ mm}$
$h = 215\text{ mm} = 215 \times 10^{-3}\text{ m}$

$$\begin{aligned}\text{so } v = \pi r^2 h &= 3.142 \times (50 \times 10^{-3})^2 \times 215 \times 10^{-3}\\ &= 3.142 \times 5^2 \times 10^{-4} \times 2.15 \times 10^{-1}\\ &= 3.142 \times 25 \times 2.15 \times 10^{-5}\\ &= 168.88 \times 10^{-5}\\ &= 1.689 \times 10^{-3}\text{ m}^3 \quad \textit{Ans}\end{aligned}$$

and $m = \rho v$

$$\begin{aligned}&= 8.5 \times 10^3 \times 1.689 \times 10^{-3}\\ &= 14.36\text{ kg} \quad \textit{Ans}\end{aligned}$$

Rough check: $v \approx 3 \times (5 \times 10^{-2})^2 \times 0.2 = 15 \times 10^{-4} = 1.5 \times 10^{-3}\text{ m}^3$

$m = \rho v \approx 9 \times 10^3 \times 1.5 \times 10^{-3} = 13.5\text{ kg}$

so our approximate results check with the 'accurate' ones.

2 The time for one complete oscillation (there-and-back swing) of a simple pendulum is given by the formula

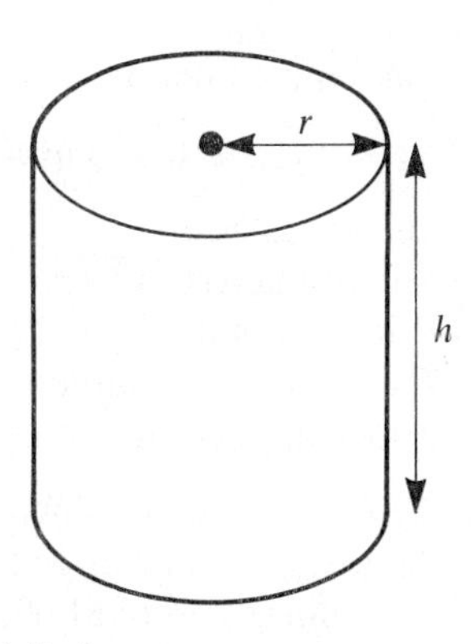

Figure 9.1 Cylinder.

$$T = 2\pi\sqrt{\frac{l}{g}} \quad \text{seconds}$$

where l = length of pendulum in metres, m
$g = 9.81\ \text{m/s}^2$, the acceleration due to gravity
$\pi = 3.142$ (to 3 decimal places)

We can also change the subject of the formula and express the length l in terms of T and g. This gives,

$$l = \frac{gT^2}{4\pi^2} \quad \text{metres}$$

Using the above formulae, calculate the periodic time T for a pendulum of length 6 inches (1 inch = 25.4 mm) and also the length of a pendulum that executes one oscillation per second.

Solution

Before substituting into the formula for T we must first convert the 6 inch length into metres,

$$l = 6 \times 25.4 = 152.4\ \text{mm} = 0.1524\ \text{m}$$

then $T = 2 \times 3.142 \times \sqrt{\dfrac{0.1524}{9.81}}$

$$= 6.284 \times \sqrt{0.01554} = 6.284 \times 0.1246$$
$$= 0.783 \quad \text{second} \ \textit{ans}$$

One oscillation per second corresponds to $T = 1$ second. Thus on substituting $T = 1$ into the formula for l we have,

$$l = \frac{9.81 \times 1^2}{4 \times 3.142^2} = 0.248\ \text{m} = 248\ \text{mm} \quad \textit{ans}$$

3 The formulae for simple and compound interest calculations are given by:

$$I = \frac{PRT}{100} \quad \text{for simple interest}$$
$$A = P + I = (1 + R/100)^T \quad \text{for compound interest}$$

where I = interest earned
R = rate of interest in % per annum
P = principal, sum originally invested
$A = P + I$ = total amount accrued, principal plus total interest

Evaluate these formulae to find the interest difference between simple and compounded interest when a lump sum of £1,000 is invested for 20 years at an interest rate of 9%.

Solution

Simple interest, $I = \dfrac{1000 \times 9 \times 20}{100} = £1{,}800$

Compound interest,

$$A = 1000 + I = 1000(1 + 9/100)^{20}$$
$$= 1000(1.09)^{20} = 1000 \times 5.6044$$
$$= £5{,}604.40$$

hence $I = £5{,}604.40 - £1{,}000 = £4{,}604.40$ and therefore the interest difference between compound and simple interest is

$$£4{,}604.40 - £1{,}800 = £2{,}804.40 \quad \textit{Ans}$$

4 The gravitational force of attraction between two bodies is given by the formula:

$$F = \frac{Gm_1m_2}{r^2} \quad \text{newtons}$$

where m_1, m_2 = masses of the two bodies in kilograms, kg
r = distance between two bodies in metres, m
$G = 6.668 \times 10^{-11}$ is a constant known as the gravitational constant.

Calculate the force of attraction between the earth and the moon, given that the masses of the earth and moon are 5.98×10^{24} kg and 7.35×10^{22} kg, respectively. The distance between the earth and the moon is 3.85×10^8 m.

Solution

On substituting,

$m_1 = 5.98 \times 10^{24}$ for the earth's mass
$m_2 = 7.35 \times 10^{22}$ for the moon's mass
$r = 3.85 \times 10^8$ for the distance between earth and moon

we have,

$$F = \frac{6.668 \times 10^{-11} \times 5.98 \times 10^{24} \times 7.35 \times 10^{22}}{(3.85 \times 10^8)^2}$$
$$= \frac{6.668 \times 5.98 \times 7.35 \times 10^{-11+24+22}}{3.85^2 \times 10^{16}}$$
$$= 19.77 \times 10^{35-16} = 19.77 \times 10^{19}$$
$$= 1.977 \times 10^{20} \ \text{newtons} \quad \textit{Ans}$$

9.2 Transformation to change the subject of a formula

A formula is normally expressed with a given quantity as its subject. Frequently we may wish to change the formula around to make another quantity the subject.

In transposing a formula to make another quantity its subject, we must, of course, follow the rules of algebra. We regard the formula as an equation – as indeed it is – and manipulate the equation to bring the wanted quantity to one side and take the other quantities to the other side. The process is illustrated in the following examples.

Examples

1 Make z the subject of the formula:

$$x = 5z + y - 3$$

Solution

Essentially we act on the equation (formula) given above to bring all non-z terms to one side of the equation thereby leaving the z term by itself on the other. So subtracting y from both sides, we have

$$x - y = 5z + y - 3 - y$$

and then adding 3 to both sides,

$$x - y + 3 = 5z - 3 + 3$$
$$\text{hence } 5z = x - y + 3$$

and finally dividing both sides by 5, we obtain

$$z = \frac{1}{5}(x - y + 3) \quad Ans$$

2 Ohm's law relating voltage V, resistance R and current I is normally stated as,

$$V = RI$$

To make I the subject, divide both sides by R:

$$\frac{V}{R} = \frac{RI}{R}$$

$$\text{so } I = \frac{V}{R}$$

Likewise to make R the subject, divide both sides of $V = RI$ by I:

$$\frac{V}{I} = \frac{RI}{I}, \qquad \text{so } R = \frac{V}{I}$$

3 Make x the subject of the formula

$$y = \frac{(x+3)}{(2x+5)}$$

Solution

First multiply both sides by $(2x + 5)$ to remove the algebraic fraction,

$$y(2x+5) = \frac{(x+3)}{(2x+5)} \times (2x+5)$$

and then on removing brackets

$$2yx + 5y = x + 3$$

Now transfer all terms containing x to one side, say LHS, and all non-x terms to the RHS,

$$2yx - x = 3 - 5y$$

i.e. subtract x and $5y$ from both sides. Now factorize the LHS, it has x as a common factor,

$$x(2y - 1) = 3 - 5y$$

and finally to obtain x divide both sides by $(2y - 1)$,

$$x = \frac{3 - 5y}{2y - 1} \quad Ans$$

4 The length of many solid substances varies with temperature according to the formula,

$$l_2 = l_1[1 + \alpha(T_2 - T_1)]$$

where l_2, l_1 = lengths at temperatures T_1, T_2

α = a constant, known as the coefficient of linear expansion

Make α the subject of the formula and hence determine α for a material which has a length of 300 mm at 20°C which increases to 300.2 mm when its temperature is raised to 60°C.

Solution

To make α the subject,

$l_2 = l_1 + \alpha l_1(T_2 - T_1)$ on removing outer [ ] brackets

$l_2 - l_1 = \alpha l_1(T_2 - T_1)$ on subtracting l_1 from both sides

$$\frac{l_2 - l_1}{l_1(T_2 - T_1)} = \alpha \text{ on dividing both sides by } l_1(T_2 - T_1)$$

i.e.

$$\alpha = \frac{l_2 - l_1}{l_1(T_2 - T_1)} \quad Ans$$

On substituting the given data,

$$\alpha = \frac{300.2 - 300}{300(60 - 20)}$$

$$= \frac{0.2}{300 \times 40} = 1.67 \times 10^{-5} \quad Ans$$

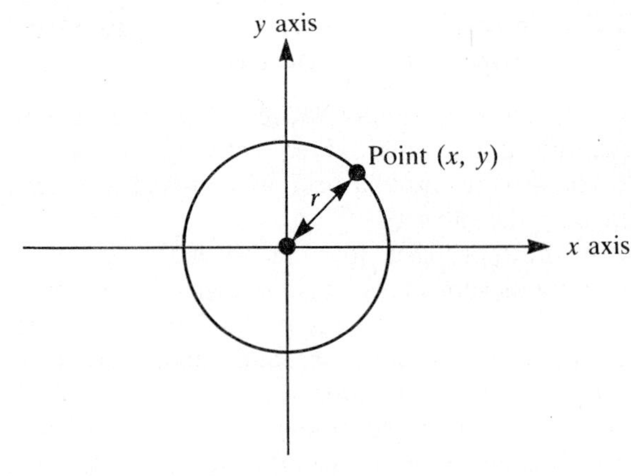

Figure 9.2 Equation of circle; $x^2 + y^2 = r^2$.

9.3 Transformations involving algebraic expressions containing indices (powers and roots)

Here again we follow the rules of algebra and apply the laws of indices to 'isolate' the required subject of the formula. The processes are illustrated in the following examples.

Examples

1 The periodic time for a simple pendulum is given by

$$T = 2\pi \sqrt{\frac{l}{g}} \text{ seconds}$$

where l = pendulum length, $g = 9.81 \text{ m/s}^2$
Express the length l as the subject of the formula.

Solution

First square both sides of the above formula to remove the square root,

$$T^2 = (2\pi)^2 \left(\frac{l}{g}\right) = 4\pi^2 \frac{l}{g}$$

then on dividing both sides by $4\pi^2$,

$$\frac{T^2}{4\pi^2} = \frac{l}{g}$$

and finally multiplying both sides by g,

$$\frac{gT^2}{4\pi^2} = l, \qquad \text{i.e. } l = \frac{gT^2}{4\pi^2} \quad Ans$$

2 The equation of a circle (see Figure 9.2), can be written as

$$x^2 + y^2 = r^2$$

where x, y define the coordinates of any point on the circle

r = radius of circle

Make y the 'subject' of this equation and hence determine its value for the case $x = 5$, $r = 13$

Solution

On subtracting x^2 from both sides of the equation, we have

$$y^2 = r^2 - x^2$$

and then on taking square roots of both sides,

$$\sqrt{y^2} = y = \sqrt{(r^2 - x^2)} \quad Ans$$

When $x = 5$, $r = 13$,

$$y = \sqrt{(13^2 - 5^2)}$$
$$= \sqrt{(169 - 25)} = \sqrt{144} = \pm 12 \quad Ans$$

3 The relationship between the distance of an object u from a lens, the distance of its image v formed by the lens, and the focal length f of a lens, see Figure 9.3, is given by

$$\frac{1}{u} + \frac{1}{v} = \frac{1}{f}$$

Express f in terms of u and v and determine f for a lens which forms an image 30 cm from the lens of an object situated 60 cm in front of the lens.

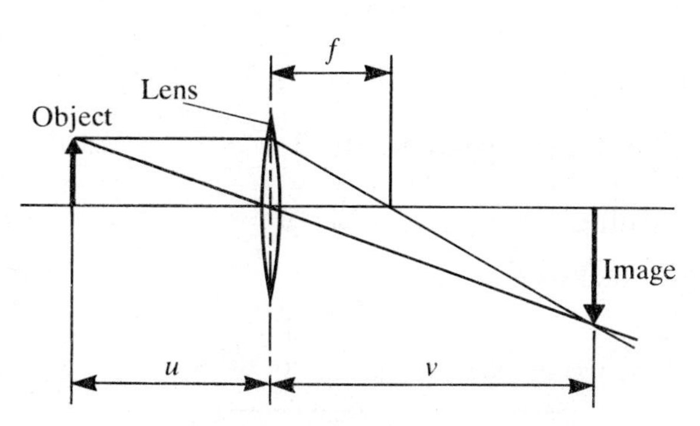

Figure 9.3 Equation relating object and image distances for a lens: $\frac{1}{u}+\frac{1}{v}=\frac{1}{f}$, *f = focal length of lens.*

Solution

First multiply the equation throughout by uv to remove the algebraic fractions on the LHS,

$$\frac{uv}{u}+\frac{uv}{v}=\frac{uv}{f}, \quad \text{i.e. } v+u=\frac{uv}{f}$$

Next invert both sides, so

$$\frac{1}{v+u}=\frac{f}{uv}$$

Finally multiply throughout by uv,

$$\frac{uv}{v+u}=\frac{fuv}{uv}, \quad \text{i.e. } f=\frac{uv}{v+u} \quad \textit{Ans}$$

When $u = 60$ cm, $v = 30$ cm,

$$f=\frac{60\times 30}{30+60}=\frac{1800}{90}=20\text{ cm} \quad \textit{Ans}$$

4 In certain types of expansion gases obey the law:

$$pv^{\gamma}=k$$

where p = gas pressure
v = volume of gas
γ = a constant, k = constant
Express v in terms of p, γ and k. If $p = 200$, $k = 1000$ and $\gamma = 1.41$, evaluate v.

Solution

Dividing both sides of the equation by p, we have

$$v^{\gamma}=\frac{k}{p} \qquad (1)$$

and on taking the γth root of both sides,

$$\sqrt[\gamma]{v^{\gamma}}=v=\sqrt[\gamma]{\frac{k}{p}} \quad \textit{Ans}$$

or equivalent raising both sides of equation (1) to the power $1/\gamma$

$$(v^{\gamma})^{1/\gamma}=v^{\gamma\times 1/\gamma}$$

$$=v=\left(\frac{k}{p}\right)^{1/\gamma} \quad \textit{Ans}$$

To evaluate v when $p = 200$, $k = 1000$, $\gamma = 1.41$ we substitute in the data

$$v=\sqrt[1.41]{\left(\frac{1000}{200}\right)}=\sqrt[1.41]{5}$$

To evaluate the 1.41th root of 5 use the $\sqrt[x]{y}$ key on an electronic calculator, i.e.

Enter 5
Press $\sqrt[x]{y}$ key
Enter 1.41
Press = key ... 3.1313 is displaced

i.e. $v=\sqrt[1.41]{5}=3.1313$ *Ans*

Alternatively, we can evaluate

$$v=\sqrt[\gamma]{5}=5^{1/\gamma} \quad \text{where } \gamma=1.41,$$
$$1/\gamma=0.7092$$

Using log tables,

$$\log v=\log 5^{0.7092}=0.7092\log 5$$
$$=0.7092\times 0.6990$$
$$=0.4957$$

so $v = \text{antilog}(0.4957)$ or $10^{0.4957}=3.131$ *Ans*

5 Express (a) R and (b) T in the compound interest formula

$$A=P(1+R/100)^{T}$$

in terms of the other variables.

Solution

(a) On dividing both sides of the formula by P, we obtain

$$\frac{A}{P}=(1+R/100)^{T}$$

Next raise both sides of the formula to the power $1/T$,

$$\left(\frac{A}{P}\right)^{1/T} = (1+R/100)^{T\times 1/T} = \left(1+\frac{R}{100}\right)^{1}$$

$$\text{so } 1+\frac{R}{100} = \left(\frac{A}{P}\right)^{1/T}$$

$$\frac{R}{100} = \left(\frac{A}{P}\right)^{1/T} - 1$$

$$R = 100\left\{\left(\frac{A}{P}\right)^{1/T} - 1\right\} \quad \textit{Ans}$$

(b) The first step to find T is to take the log of both sides of the compound interest formula. This is normally always the first step when we are solving for indices in an equation or formula.

$$A = P(1+R/100)^T$$

$$\text{so } \log A = \log P + \log(1+R/100)^T$$

$$= \log P + T\log(1+R/100)$$

$$\text{Thus } T\log(1+R/100) = \log A - \log P$$

$$T = \frac{\log A - \log P}{\log(1+R/100)}$$

Ans

6 Using the results of example 5, determine
   (a) The rate of interest $R\%$ per annum required to double a sum invested over a period of 8 years.
   (b) The time taken for a sum of £50,000 to accumulate to £150,000 at an annual rate of interest $R = 6\%$.

Solution

(a) If a sum P is to double then the amount $A = 2P$, so on substituting this into the formula, with $T = 8$ we have

$$R = 100\left\{\left(\frac{2P}{P}\right)^{\frac{1}{8}} - 1\right\}$$

$$= 100(2^{\frac{1}{8}} - 1)$$

$$= 100(1.0905 - 1),$$

$$(\text{as } 2^{\frac{1}{8}} = 2^{0.125} = 1.0905)$$

$$= 9.05\% \quad \textit{Ans}$$

(b) Substituting $A = 150{,}000$, $P = 50{,}000$ and $R = 6$ into

$$T = \frac{\log A - \log P}{\log\left(1+\frac{R}{100}\right)}$$

$$= \frac{\log 150{,}000 - \log 50{,}000}{\log(1+0.09)}$$

$$= \frac{5.1761 - 4.6990}{0.0374} = 12.76 \text{ years} \quad \textit{Ans}$$

Test and problems 9

Multiple choice test MT 9

Answer block:

Question No.	0	1	2	3	4	5	6	7	8
Answer	c								

Enter your answer, that is a, b, c or d in the column under the question number in the answer block above. Note that question Qu. 0 has already been worked out and the answer inserted.

Qu. 0
Make l the subject of the formula
$S = 2\pi r^2 + 2\pi rl$.
Ans (a) $S - 2\pi r^2 - 2\pi r$ (b) $S/2\pi r^2 + 2\pi r$
(c) $(S - 2\pi r^2)/2\pi r$ (d) $S/2\pi r - 2\pi r^2$

Solution

Subtracting $2\pi r^2$ from both sides,

$$S - 2\pi r^2 = 2\pi rl$$

and dividing throughout by $2\pi r$,

$$\frac{S - 2\pi r^2}{2\pi r} = l$$

so (c) corresponds to the correct answer, hence 'c' is inserted under Qu. 0 in the answer block as shown

Now carry on with the test.

Qu. 1 The kinetic energy E of a body of mass m kg moving at a speed of v m/s is given by $E = \frac{1}{2}mv^2$ joules.
Evaluate E for $m = 10$ kg, $v = 20$ m/s
Ans (a) 200 (b) 2000 (c) 4000
(d) 500

Qu. 2 Calculate R in the formula $R = \dfrac{R_1 R_2}{R_1 + R_2}$ for the case $R_1 = 600$, $R_2 = 1200$

Ans (a) 900 (b) 720 (c) 1800 (d) 400

Qu. 3 Make m the subject of the formula $v = \sqrt{\dfrac{T}{m}}$

Ans (a) $v^2 T$ (b) v^2/T (c) T/v^2 (d) $\sqrt{(T/v)}$

Qu. 4 Determine a in the formula $v^2 = u^2 + 2as$ for case $v = 10$, $u = 4$, $s = 5$

Ans (a) 8.4 (b) 0.6 (c) 3.6 (d) 11.8

Qu. 5 Make β the subject of the formula $R = R_0(1 + \beta T)$

Ans (a) $(R - R_0)/T$ (b) $R/R_0 - 1/T$ (c) $(R - R_0)/R_0 T$ (d) $R/R_0 T$

Qu. 6 Make b the subject of the formula $v = 3.14(b^2 - a^2)$

Ans (a) $v^2/3.14 + a^2$ (b) $\sqrt{(v/3.14 + a^2)}$ (c) $v/3.14a^2$ (d) $(v/3.14 + a^2)^2$

Qu. 7 Evaluate $A = P(1 + R/100)^T$ for the case $P = 500$, $R = 10$, $T = 4$.

Ans (a) 732.05 (b) 605 (c) 2200 (d) 700

Qu. 8 Evaluate v using the formula $\dfrac{1}{u} + \dfrac{1}{v} = \dfrac{1}{f}$ for the case $u = 50$, $f = 10$

Ans (a) 40 (b) 12.5 (c) 60 (d) 15

Problems 9

1 Calculate the value of R in the formulae

(a) $R = \dfrac{R_1 R_2}{R_1 + R_2}$ when $R_1 = 1200$, $R_2 = 600$

(b) $R = R_1 + R_2 + R_3 R_4/(R_3 + R_4)$ when $R_1 = 1000$, $R_2 = 3300$, $R_3 = 6000$, $R_4 = 30{,}000$

2 Evaluate P to two decimal places using the formula

$P = RI^2$ when $R = 9300$ and $I = 0.0129$

3 The following data gives values for the conversion from Imperial to metric units for:

1 mile = 1609 metres
1 gallon = 4.546 litres,
8 pints = 1 gallon
1 lb = 0.4536 kilograms

Write formula defining the terms used for

(a) converting miles to kilometres;
(b) converting litres to pints;
(c) converting pounds (lb) to kilograms and vice-versa.

4 Transpose the following formulae to change the subject.

(a) $v = u + at$, make a the subject
(b) $y = mx + c$, make c the subject
(c) $v = \dfrac{4}{3}\pi r^3$, make r the subject
(d) $f\lambda = c$, make λ the subject
(e) $v = \pi r^2 h$, make r the subject
(f) $I = PRT/100$, make R the subject
(g) $\omega = 1/\sqrt{(LC)}$, make C the subject
(h) $Z = \sqrt{(R^2 + X^2)}$, make X the subject

5 The resistance to electrical current for many materials is given by the formula,

$$R = \rho \frac{l}{A} \text{ ohms}$$

where l = length of conductor material in metres
A = cross-sectional area of conductor in square metres
ρ = resistivity of material

Calculate:

(a) the resistance of a coil consisting of 500 m of wire of cross-sectional area 0.665 mm^2 and resistivity 1.56×10^{-8}
(b) the resistance of wire of radius 1 mm drawn out from a melt of 1 cubic metre of metal of resistivity 2.8×10^{-7}.
Note: cross-sectional area of circle $= \pi r^2$
volume of length l of wire of radius $r = \pi r^2 l$,
take $\pi = 3.142$

6 The velocity of sound at sea level is given by the formula,

$v = 331.46 + 0.61\ T$ metres per second

where T = temperature in °C.

The general formula for the velocity of waves also applies,

$$v = f\lambda$$

where f = frequency of source emitting waves, units hertz (Hz)

λ = wavelength of waves, units metres (m).

Using the above formulae, determine

(a) the frequency of a tuning fork which produces a soundwave of wavelength 1.294 m at 0°C;

(b) the wavelength of the sound-waves produced by a source of frequency of 1 kHz when the temperature is 30°C.

10 Direct and inverse proportionality

General learning objectives: to illustrate direct and inverse proportionality.

10.1 Dependent and independent variables

We saw in Chapter 9 the use of formulae to express the subject in terms of other quantities. We also transformed formulae to change to new subjects. In general a formula or an equation shows how quantities – the variables – depend on one another.

For example, the equation $y = 2x + 7$ shows how the quantity or variable y depends on the variable x. The quantity y which depends on x is known as the **dependent variable** in the above equation. The quantity x which determines the value of y is known as the **independent variable**.

In plotting graphs of an equation the independent variable is normally plotted on the horizontal or x-axis. The dependent variable values calculated from a series of x values is plotted on the vertical or y-axis.

Examples

1 Ohm's law expressed as voltage = resistance × current

$$V = RI$$

has the voltage V as the dependent variable and the current I as the independent variable.

The resistance R is usually a constant – its value does not vary – and the law expresses the physically observed fact that V *depends* on I or more specifically that V is directly proportional to I.

2 The periodic time for a simple pendulum is given by

$$T = 2\pi\sqrt{\frac{l}{9.81}}$$

T is the dependent variable and the pendulum length l is the independent variable. T *depends* on the square root of l.

10.2 Proportionality statements: direct and inverse proportionality

It is common practice to express the dependence of related quantities as statements of proportionality. For example, Ohm's law relating voltage and current, see Figure 10.1(a), may be stated as:

> the voltage developed across the terminals of a resistor *is directly proportional to* the current flowing in the resistor.

This may be written as the proportional statement:

$$V \propto I$$

where the proportionality symbol $\propto$ means *is directly proportional to* or *varies directly as*; i.e. if, for example, the current is doubled then the voltage doubles; if the current is decreased by a factor of 3, then so is the voltage.

Hooke's law, see Figure 10.1(b) may be expressed as:

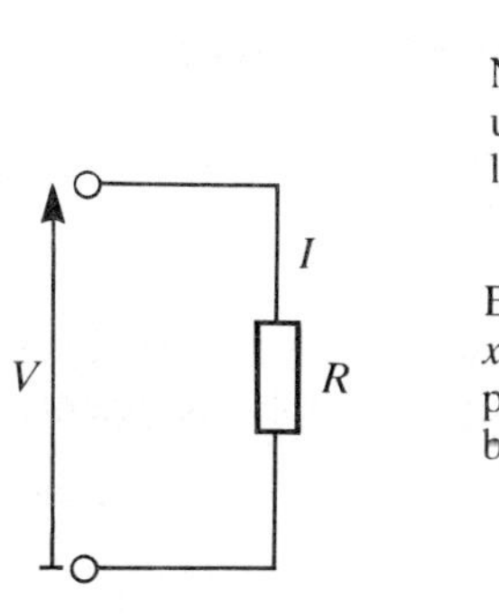

(a) Ohm's law: voltage $\propto$ current $V \propto I$

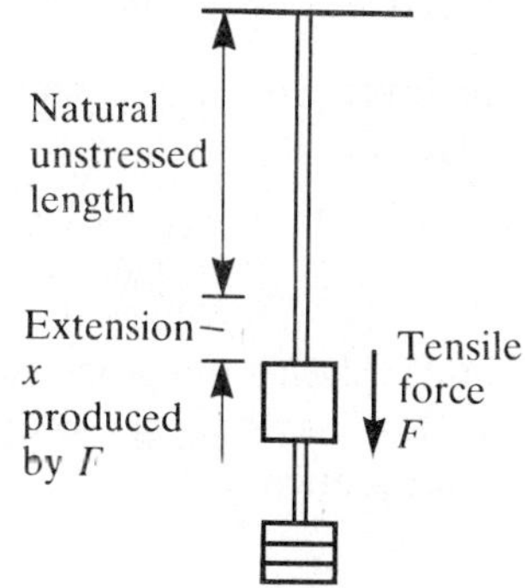

(b) Hooke's law: force $\propto$ extension $F \propto x$

Figure 10.1.

the tensile force acting on a material *is directly proportional to* the extension it produces, i.e.

$$F \propto x$$

Proportionality statements can be converted into an equation by replacing the $\propto$ symbol by the equals sign and including a constant multiplier, e.g.

For Ohm's law: $V \propto I$ becomes

$$V = RI$$

where in this case the constant of proportionality is the resistance of the resistor, R. If, for example a current of 5 amperes produces a voltage of 100 volts, then

$$R = \frac{100}{5} = 20 \text{ Ohms}$$

For Hooke's law: $F \propto x$ may be written as

$$F = \lambda x$$

where the constant of proportionality λ could be determined experimentally. For example:

If a force $F = 200$ newtons produced an extension of 1 mm ($= 10^{-3}$ m) in a specimen of mild steel wire, then for this particular wire:

$$\lambda = \frac{F}{x} = \frac{200}{10^{-3}} = 200 \times 10^3 = 2 \times 10^5$$

As well as 'directly' proportional statements, statements of 'inverse' proportionality are often encountered. The statements: '*y is inversely proportional to*, and '*y varies inversely as x*' are examples of how inverse proportionality statements are expressed in words and means that the dependent variable y is proportional to $\frac{1}{x}$ (the reciprocal or inverse of x). The statement expressed mathematically is

$$y \propto \frac{1}{x}$$

and replacing the proportionality symbol $\propto$ by $=$, we express the inverse proportionality statement as the equation

$$y = \frac{k}{x}$$

where k is the constant of proportionality. For example, Boyle's law for gases, see Figure 10.2, may be stated as:

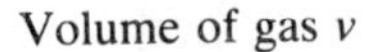

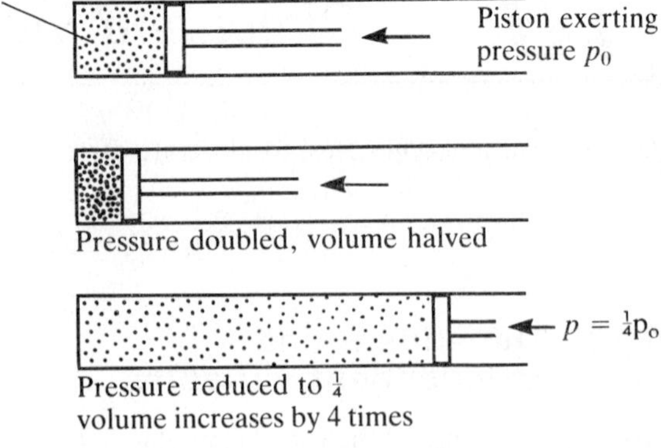

Figure 10.2 Boyle's law: volume $\propto \frac{1}{pressure}$, $v \propto 1$ *for a gas at constant temperature.*

The volume of a fixed mass of gas maintained at a constant temperature *is inversely proportional to* the pressure exerted on the gas.

This statement may be written as

$$v \propto \frac{1}{p}$$

Thus, for example, if the pressure p on the gas is doubled ($\times 2$) the volume v is halved $\left(\times \frac{1}{2}\right)$; if the pressure is reduced to $\frac{1}{4}$ of its value, the volume expands by $1 \Big/ \frac{1}{4} = 4$ times.

A second example of inverse proportionality is the 'inverse square' law in radiation. The power density due to a source of radiation such as the sun or a radio transmitter varies inversely as the square of the distance. This fact may be written as

$$P \propto \frac{1}{r^2}$$

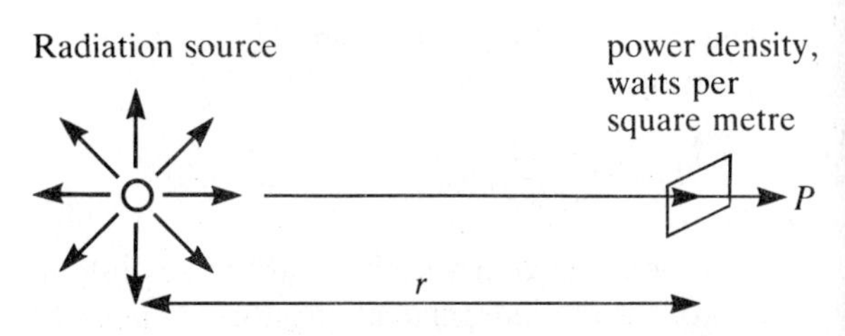

Figure 10.3 Inverse square law, $P \frac{1}{r^2}$.

where P = power density in watts per square metre, W/m^2; r = distance from source of radiation, see Figure 10.3

Direct and inverse proportionality statements can be combined. For example, the force of repulsion between two similarly charged bodies is directly proportional to the product of the charges on the body but is inversely proportional to the square of the distance between them. The individual proportionality statements may be written as:

$F \propto Q_1 Q_2$, i.e. force F is directly proportional to product of the charges Q_1 and Q_2.

$F \propto \frac{1}{d^2}$, i.e. force F is inversely proportional to the square of the distance d.

The two statements are combined to form a single statement by multiplying:

$$F \propto Q_1 Q_2 \times \frac{1}{d^2}, \quad \text{i.e.} \quad F \propto \frac{Q_1 Q_2}{d^2}$$

10.3 Practical applications of proportionality

Proportionality statements such as

$y \propto x$	direct proportionality
$y \propto \frac{1}{x}$	inverse proportionality
$F \propto \frac{Q_1 Q_2}{d^2}$	F directly proportional to $Q_1 Q_2$ and inversely proportional to d^2

can always be converted into equations by replacing the proportional sign $\propto$ by the equals sign = and including a constant multiplier, i.e.

$y \propto x$ is exactly equivalent to $y = k_1 x$

$y \propto \frac{1}{x}$ is equivalent to $y = \frac{k_2}{x}$ or $xy = k_2$

$F \propto \frac{Q_1 Q_2}{d^2}$ can be expressed as $F = k\frac{Q_1 Q_2}{d^2}$

where the constants k_1, k_2, k are known as *coefficients or constants of proportionality*. These constants may be determined by substituting known values for the variables into the equations.

Examples

1 It is known that y is directly proportional to x and the following values for x and y are determined by experiment:

x	0	1	2	3	4
y	0	2.2	4.5	6.6	8.8

Write down the equation relating y to x.

Solution

As $y \propto x$, the equation relating y to x is

$$y = kx$$

where k = coefficient or constant of proportionality. k may be evaluated using the above table of values:

when $x = 0$, $y = 0$ which is obviously satisfied by $y = kx$
when $x = 1$, $y = 2.2$ so $2.2 = k \times 1$, $k = 2.2$
when $x = 2$, $y = 4.5$ so $4.5 = k \times 2$, $k = 4.5/2 = 2.25$
when $x = 3$, $y = 6.6$ so $6.6 = k \times 3$, $k = 6.6/3 = 2.2$
when $x = 4$, $y = 8.8$ so $8.8 = k \times 4$, $k = 8.8/4 = 2.2$

so apart from one pair of results we have $k = 2.2$ and even the result that gives $k = 2.25$ differs only in the third significant figure. Thus allowing for experimental variation, we have

$$y = 2.2x \quad \textit{Ans}$$

2 P varies inversely with the square of r. It is also known that when $r = 10$, $P = 50$. Determine P when $r = 75$

Solution

Since $P \propto \frac{1}{r^2}$, we may write $P = \frac{k}{r^2}$

and on substituting $P = 50$, $r = 10$ we have $k = Pr^2 = 50 \times 10^2 = 5000$

Thus the general equation relating P and r is

$$P = \frac{5000}{r^2}$$

so when $r = 75$, $P = \frac{5000}{75^2} = 0.8889$ *Ans*

3 Charles' law for gases states that the volume of a fixed mass of gas held under constant pressure

is directly proportional to the absolute temperature of the gas.

A fixed mass of gas held at constant pressure has a volume of $500\,cm^3$ at a temperature of 0°C (273K). Calculate the volume of the gas at -100°C and also at $+50$°C.

Note: absolute temperature is measured in units known as Kelvins (K) and is related to temperature measured in degrees Celsius (°C) by:

$$T\,K = T°C + 273\,K$$

so when $T°C = -100°C$, $T = -100 + 273 = 173K$

when $T°C = +50°C$, $T = 50 + 273 = 323K$

Solution

Let v = volume and T = absolute temperature, then as Charles' law applies we have:

$$v \propto T, \qquad v = kT$$

When $v = 500$, $T = 273K$ (0°C), so

$$500 = 273k, \qquad k = 500/273 = 1.8315$$

Hence in general for the fixed mass of gas,

$$v = 1.8315T$$

When $T = 173K$ $(-100°C)$,
$v = 1.8315 \times 173 = 316.8\,cm^3$ *Ans*

When $T = 323K$ $(+50°C)$,
$v = 1.8315 \times 323 = 591.6\,cm^3$ *Ans*

4 The deflection system in a moving-iron ammeter used for measuring alternating current has a deflection proportional to the square of the current. If a meter of this type has a full-scale deflection when measuring a current of 5A, calculate the current flowing when the deflection is registering half full-scale deflection.

Solution

Let x = meter deflection corresponding to current of I. Then, as the deflection is proportional to the square of the current,

$$x \propto I^2 \quad \text{or} \quad x = kI^2$$

using the fact that $x = x_{max}$ when $I = 5$, we have

$$x_{max} = k5^2 = 25k, \quad \text{so } k = x_{max}/25$$

When the deflection, $x = \frac{1}{2}x_{max}$:

$$\frac{1}{2}x_{max} = \frac{x_{max}}{25} \times I^2$$

so cancelling x_{max} on both sides of the equation, we obtain

$$\frac{1}{2} = \frac{I^2}{25}$$

$$I^2 = \frac{25}{2}, \quad \text{so } I = \sqrt{\frac{25}{2}} = 3.536\,A \quad \textit{Ans}$$

Test and problems 10

Multiple choice test: MT 10

Answer block:

Question No.	0	1	2	3	4	5	6	7	8
Answer	c								

Enter your answer, that is a, b, c or d in the column under the question number in the answer block above. Note that question Qu. 0 has already been worked out and the answer inserted.

Qu. 0 The volume of a gas held at constant temperature varies inversely with pressure. If the volume of a gas is $100\,cm^3$ at 20°C, determine its new volume if the pressure is reduced by a factor of 6, the temperature remaining at 20°C.
Ans (a) 6 (b) 16.7 (c) 600 (d) 1000

Solution

Since the volume v varies inversely with pressure p, we have

$$v \propto \frac{1}{p}$$

so if the pressure is reduced by a factor of 6, i.e. p becomes $\frac{1}{6}p$, the new volume,

$$v_n \propto \frac{1}{\frac{1}{6}p} = \frac{6}{p}$$

is increased by six times,

so $v_n = 6v = 6 \times 100 = 600\,\text{cm}^3$

so the correct answer is (c) and therefore c is inserted in the answer block under Qu No. 0.

Now carry on with the test.

Qu. 1 It is known that the variable y is directly proportional to x and that when $x = 10$, $y = 100$. Determine the value of y when $x = 15$

Ans (a) 50 (b) 200 (c) 115 (d) 150

Qu. 2 y is inversely proportional to x. This statement may be expressed mathematically as

Ans (a) $y \propto x$ (b) $y = kx$ (c) $y \propto \frac{1}{x}$
(d) $y = \frac{1}{x}$

Qu. 3 If y is directly proportional to x and the constant of proportionality is 5, which of the following 'answers' cannot be true.

Ans (a) $x = 1$, $y = 5$ (b) $x = 0$, $y = 0$
(c) $x = -5$, $y = -25$
(d) $x = 10$, $y = 100$

Qu. 4 Hooke's law states that the tensile force is directly proportional to the extension it produces. In a test on copper wire a force of 100 N produces an extension of 2.5 mm. Determine the extension produced by a force of 30 N.

Ans (a) 7.5 mm (b) 0.75 mm
(c) 0.5 mm (d) 1.5 mm

Qu. 5 The periodic time of a simple pendulum is directly proportional to the square root of its length. A pendulum of length 25 cm has a periodic time of just over 1 second. Determine the approximate periodic time for a pendulum of length 100 cm.

Ans (a) 4s (b) 0.5s (c) 16s (d) 2s

Qu. 6 In an electronic device the current I is proportional to the square of the voltage V impressed across the device. It is known that when $V = 0.6$ volts, $I = 0.1$ amps (A). Determine the current when $V = 0.3$ volts.

Ans (a) 0.025A (b) 0.5A (c) 0.05A
(d) 0.15A

Problems 10

1 It is known that y is directly proportional to x; determine the actual equation for the relationship for the cases when it is known that
(a) when $x = 1$, $y = 1$
(b) when $x = 9$, $y = 27$
(c) when $x = -5$, $y = 10$

2 If y varies inversely with x, write the equation relating y and x, given
(a) when $x = 0.5$, $y = 1$
(b) when $x = 4$, $y = 40$

3 The volume of a sphere is directly proportional to the cube of its radius. If the volume of a sphere of unit radius is 4.19 units determine the volume of a sphere of radius r when
(a) $r = 0.5$ (b) 2 (c) 10

4 The volume of a fixed mass of gas maintained at constant temperature varies inversely as the pressure exerted on the gas. If a gas occupies a volume of 30 litres and is maintained at the temperature of 0°C calculate the volume when the pressure is (a) increased by a factor of 5, (b) reduced by a factor of 10 from its initial value.

5 The force of attraction between two magnets is inversely proportional to the distance between the magnets. If the force equals 10 N when the magnets are 1 m apart determine the force when the magnets are separated by (a) 10 m, (b) 10 cm

6 The force between two electrical charges is directly proportional to the product of the charges and inversely proportional to the square of the distance between the charges. If the force of repulsion between two positively charged bodies situated 100 mm apart is 1 N, determine the force when
(a) the distance between the bodies is reduced to 20 mm;
(b) the distance is increased to 200 mm and the charge on each body is doubled.

11 Equation of straight-line graph

General learning objective: to determine the equation of a straight-line graph.

11.1 Cartesian coordinates: *x-y* graphs

Before we consider the equation of a straight line let us recap on some basic terms concerning graphs and see how points are plotted to form a graph.

The term 'graph' is applied normally to the line trace which defines the relationship between two variables, say, x and y. A graph is constructed by plotting points, that is values of y for various values of x on graph paper and then drawing a smooth curve through the plotted points.

The framework used to plot the points consists of two axes drawn at right angles to each other, as shown in Figure 11.1. The horizontal axis or x-axis is normally used for the independent variable, the variable which determines y. The vertical or y-axis is used for the dependent variable. Appropriate scales are marked on the axes – it is usually up to us to decide the most suitable scale for both the x and y axes so as to ensure all points of interest may

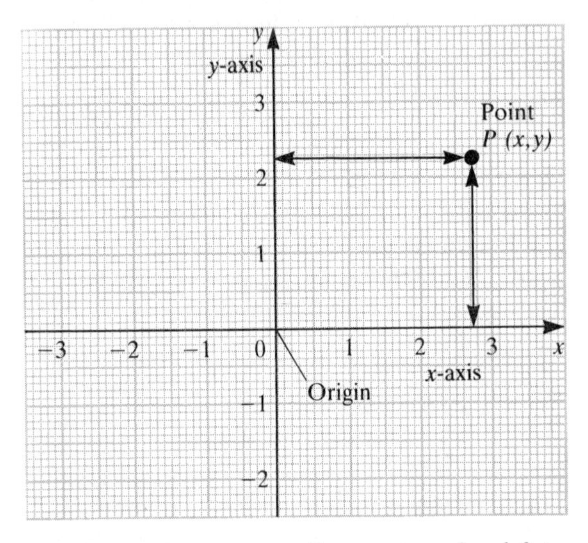

Figure 11.1 Cartesian coordinate system for defining points and plotting x–y graphs.

be plotted. The zero point, where $x=0$ and $y=0$, is at the intersection of the two axes and is known as the **origin**. x values to the right of the origin are positive, values to left of the origin are negative. y values above the x-axis are positive, and below the x-axis are negative.

A point defining the value of y for a value of x is plotted with respect to these two axes by moving y units up (if positive), down (if negative) parallel to the y-axis and then x units parallel to the x-axis, to the right of the origin if the x value is positive to the left if the value is negative. Points are usually marked on the graph paper by a dot or a small cross.

This system in which a point is located in a plane by specifying its distance from two axes drawn at right angles, i.e. from the y and x-axes, is known as the **cartesian system of coordinates**. The values of x and y defining the point are known as **cartesian coordinates**. The perpendicular distance of the point from the y axis, that is the x value is known as the **abscissa**; the perpendicular distance of the point from the x axis, that is the y value, is known as the **ordinate**. The mathematical convention used to denote the coordinates of a point is (x, y), e.g.

> if point P is defined by $x=4$, $y=-7$ it is denoted by $(4, -7)$
>
> ↑ y value (ordinate)
>
> ↑ x value (abscissa)

Example

Draw a cartesian system of axes, i.e. x and y axes, selecting an appropriate scale for both axes so that the following points may be plotted:

> A$(-6, -2)$, B$(-2, 0.67)$, C$(3, 4)$, D$(5, 5.33)$

Plot these points and show that they lie on a straight line. Determine the intercepts of this line on the x and y axes.

Solution

First select appropriate scales for the x and y axes, noting the range of x and y values for the points. x varies between -6 for point A to $+5$ for point D; y varies from -2 (point A) to $+5.33$ (point D). Thus our x axis should cover at least -6 through 0 to $+5$ and our y axis from -2 to $+5.33$. In Figure 11.2 we have used centimetre square graph paper where the 'bigger' squares are 1 cm by 1 cm and the finer detail lines are spaced by one millimetre (remember 1 cm = 10 mm). We have used the same scale on both axes of 1 unit = 1 cm.

Point A, defined by $(-6, -2)$ i.e. $x = -6$, $y = -2$, is plotted by moving -6 units along the x-axis, that is in the negative direction from the origin and then -2 units parallel to the y axis, i.e. downwards.

Point B, $(-2, 0.67)$ has $x = -2$, $y = +0.67$ and is plotted by moving -2 units along the x-axis and then 0.67 units (0.67 mm) upwards parallel to the y axis.

Point C, $(3, 4)$ has $x = 3$, $y = 4$ and is plotted by moving 3 units along the x-axis and 4 units in the positive y-axis direction.

Finally point D, $(5, 5.33)$ where $x = 5$ and $y = 5.33$ is plotted.

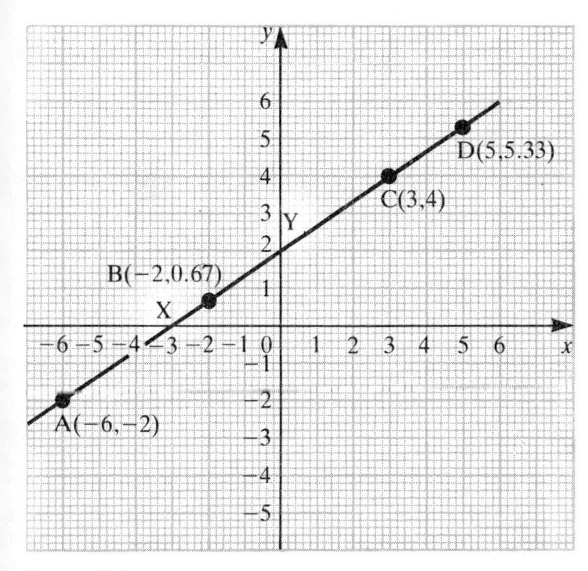

Figure 11.2.

The fact that points A, B, C and D lie on a straight line can now be checked easily by lining up a ruler on A and D and noting that the two intermediate points B and C are all in line. In Figure 11.2 the straight line has been drawn.

The intercept of this line on the y axis, i.e. the value of y where the line cuts the y axis, can now be read off:

intercept on y-axis is $y = 2$ (point Y)

Likewise the point where the line intercepts the x-axis can also be read off from the graph:

intercept on x-axis is $x = -3$ (point X)

11.2 The equation of a straight line: $y = mx + c$

Let us consider now how we can determine the equation of a straight line by reference to Figure 11.3. Here we have drawn-in a line AB and our problem is to form a relation between y and x which defines all points which lie on this line.

One important point of reference is where the line cuts the y axis, the **intercept** on the y-axis, which we denote by c. Then when $x = 0$, we have $y = c$.

Now let us consider the slope or gradient of the line. When x increases from 0 to 1, suppose y increases by an amount m, then when $x = 1$, $y = c + m$. If we increase x from 1 to 2, y increases by

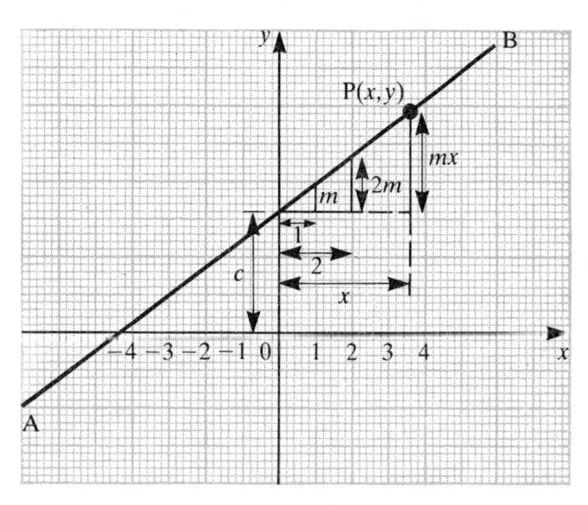

Figure 11.3 General equation of a straight line $y = mx + c$ where m = gradient, c = intercept on y-axis.

the same amount m again since the slope of a straight line does not change so we may write: when $x = 2$, $y = c + 2m$ and in general, if x is increased from 0 to any value x, y increases by $m \times x$ and so we obtain the general equation:

$$y = mx + c$$

where $c =$ intercept on y-axis (value of y when $x = 0$)
$m =$ change in y/change in x, known as the gradient of the straight line, which we consider in detail in the next section.

Although we have constructed the equation for a straight line by considering a line which 'slopes' upwards – one with a positive gradient m, the equation $y = mx + c$ is quite general and holds for all values of the gradient m, whether positive, negative or zero and likewise for all values of c. It is the characteristics of the line which determine m and c and we determine these values by substituting into the general equation known data so as to obtain the actual equation for a particular line.

Examples

1 Determine the equation for the following straight lines, given

(a) the line has an intercept of 5 on the y-axis and when x is increased from 0 to 4, y increases from 5 to 17;

(b) the line has an intercept of 2 on the y-axis and when x is increased from 0 to 4, y decreases from 2 to -6.

Solution

(a) The general equation of the straight line is given by
$y = mx + c$ where, in this case,

$c = 5$, intercept on y-axis (value when $x = 0$)

$$\text{and } m = \frac{\text{change in } y}{\text{change in } x} = \frac{17-5}{4-0} = \frac{12}{4} = 3$$

Hence $y = 3x + 5$ *Ans*
The line is also plotted Figure 11.4(a)

(b) In this case $c = 2$, and

$$m = \frac{\text{change in } y}{\text{change in } x} = \frac{-6-(+2)}{4-0}$$

$$= \frac{-8}{4} = -2$$

Note: in this case the line 'slopes' downwards since y decreases as x increases and m is therefore negative. Substituting $m = -2$ and $c = 2$ into the general equation, $y = mx + c$, we obtain

$$y = -2x + 2 \quad Ans$$

This line showing the negative gradient characteristic is plotted in Figure 11.4(b).

2 Plot the straight line passing through the points $(-1, -5)$ and $(5, 7)$ and check that its intercept on the y-axis is at $y = -3$ and the intercept on the x-axis is at $x = 1.5$. Determine also the equation of the line.

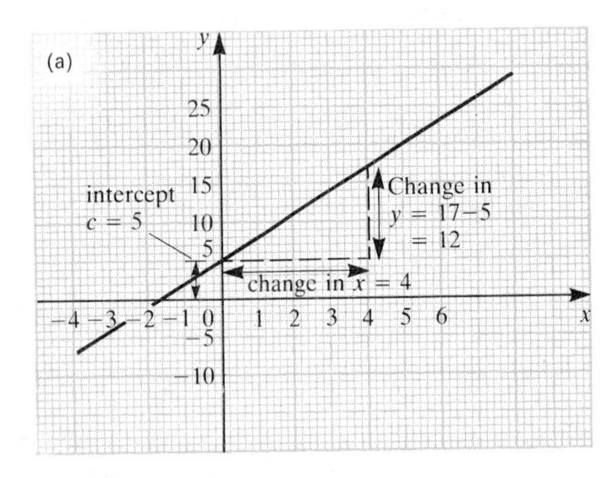

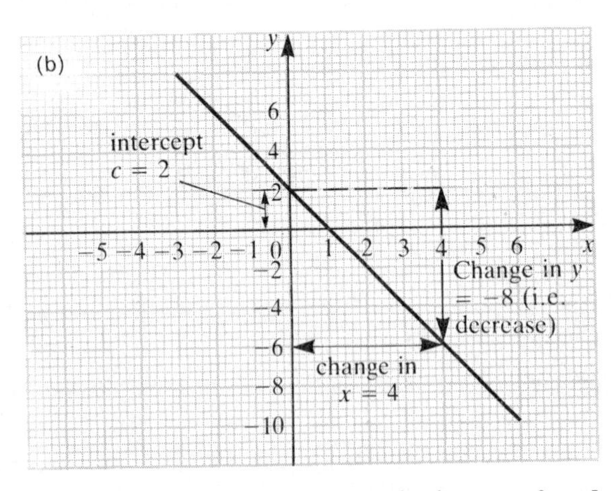

Figure 11.4 (a) Plot of a straight line: $y = 3x + 5$ (b) Plot of a straight line $y = 2\bar{x} + 2$.

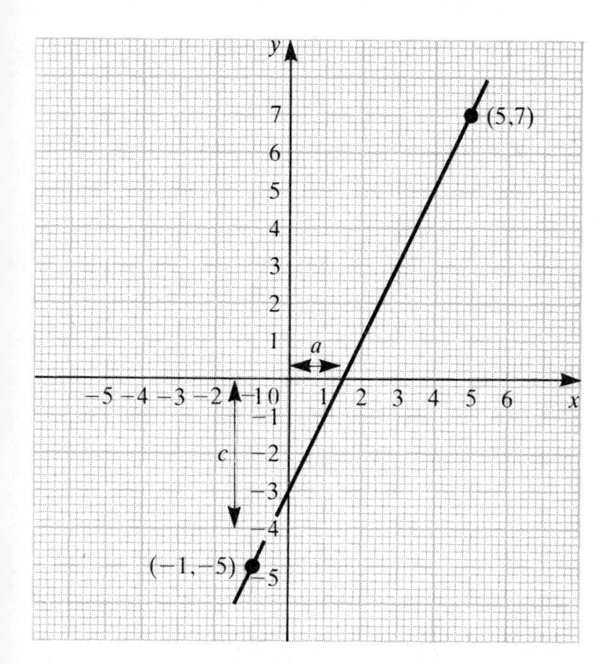

Figure 11.5 Plot of a straight line passing through the points $(-1, -5)$ and $(5, 7)$: equation $y = 2x - 3$.

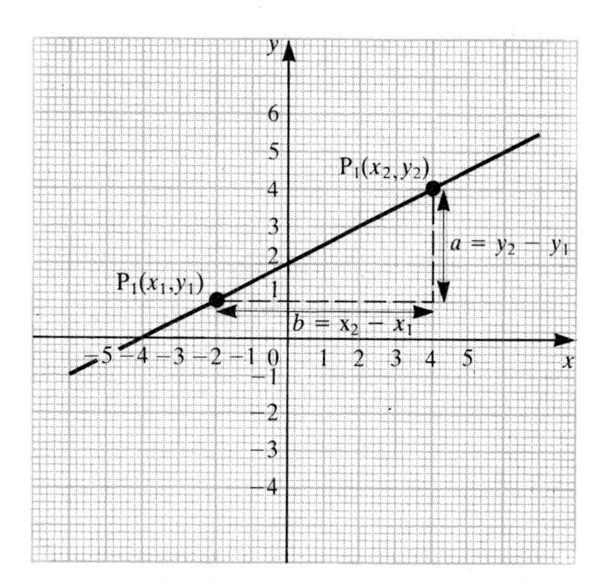

Figure 11.6 Definition of gradient m of a straight line:

$$m = \frac{\textit{change in y value}}{\textit{change in x value}}$$

Solution

The two points $(-1, -5)$ and $(5, 7)$ are plotted in Figure 11.5 and a straight line drawn through them. Reading from the graph, we have

intercept on y-axis, $c = -3$ *Ans*
intercept on x-axis, $a = 1.5$ *Ans*

Using the general equation for the straight line, we have

$$y = mx + c = mx - 3, \quad \text{as} \quad c = -3.$$

m can be determined by substituting the coordinates of any point on the line, so, for example if we use the point $(5, 7)$, i.e. $x = 5$, $y = 7$ we obtain,

$$7 = 5m - 3$$

hence $5m = 7 + 3 = 10$, $m = 2$ and the equation of the line is

$$y = 2x - 3 \quad \textit{Ans}$$

Alternatively we can calculate m by finding the gradient of the line,

$$m = \frac{\text{change in } y}{\text{change in } x} = \frac{7 - (-5)}{5 - (-1)} = \frac{12}{6} = 2$$

11.3 The gradient of a straight-line graph

The gradient of a straight line as we have already seen is a measure of the inclination or slope of the line and relates how y changes as x changes.

Mathematically we define the gradient of a straight line as the ratio of the change in the value of y (the dependent variable) to the change in the value of x (the independent variable) between any two points which lie on the straight line. Thus, referring to Figure 11.6, the gradient of the straight line is defined as:

$$m = \frac{\text{change in } y \text{ value between points } P_1 \text{ and } P_2}{\text{change in } x \text{ value between same points}}$$

$$= \frac{a}{b}$$

If the coordinates of the two points P_1 and P_2 are respectively (x_1, y_1) and (x_2, y_2), then

change in y, $a = y_2 - y_1$
change in x, $b = x_2 - x_1$

so

$$m = \frac{y_2 - y_1}{x_2 - x_1}$$

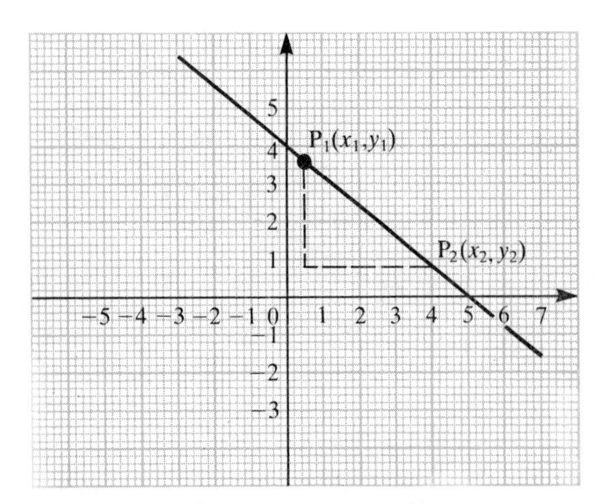

Figure 11.7 Line with negative gradient.

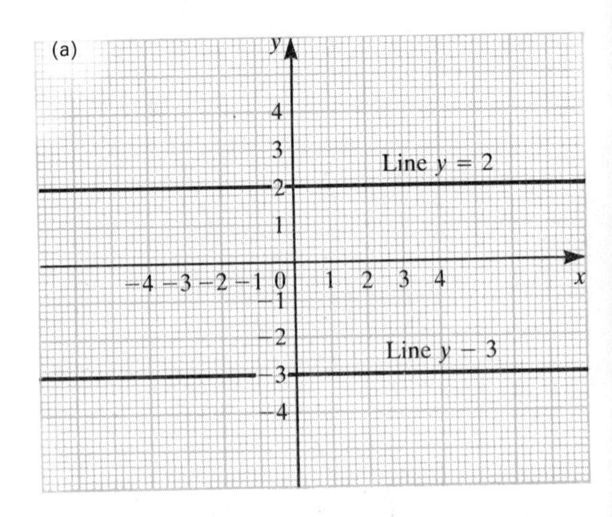

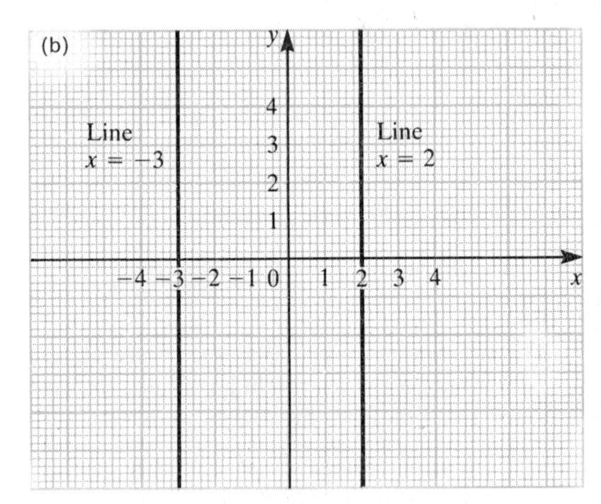

Figure 11.8 (a) Lines parallel to x-axis have zero gradient, m = 0 (b) Lines parallel to y-axis.

For example if P_1 is $(-2, 1)$ and P_2 is $(4, 4)$ then

$$m = \frac{4-1}{4-(-2)} = \frac{3}{4+2} = \frac{3}{6} = \frac{1}{2}$$

Note: wherever the points P_1 and P_2 are selected on the line the gradient will be the same.

The gradient of a straight line represents the rate at which y changes relative to x. If y increases with increasing x then the gradient is a positive value and the line makes an acute angle (angle less than 90°) with the x-axis. If, however, y decreases with increasing x, then the gradient of the line will have a negative value and make an obtuse angle (angle between 90° and 180°) with the x-axis. For example, in Figure 11.7, the gradient of the line,

$$m = \frac{y_2 - y_1}{x_2 - x_1}$$

is clearly negative as y_2 is less than y_1. For the points selected on the line, P_1 is $(0.5, 3.6)$ and P_2 is $(4, 0.8)$, so

$$m = \frac{0.8 - 3.6}{4 - 0.5} = \frac{-2.8}{3.5} = -0.8$$

Thus, in general, lines which rise from left to right have a positive gradient ($m > 0$); lines that fall from left to right have a negative gradient ($m < 0$). In between there are two special cases: lines which are parallel to the x-axis and lines which are parallel to the y-axis.

Lines parallel to the x-axis, see Figure 11.8(a), have a zero gradient, so $m = 0$. There is no change in the value of y as x is changed, y remains constant. The general equation for such lines is:

$$y = \text{constant}$$

and shown in Figure 11.8(a) are two examples,

$$y = 2 \quad \text{and} \quad y = -3$$

Lines parallel to the y-axis have an infinitely large gradient and can be defined by the general equation,

$$x = \text{constant}$$

and shown in Figure 11.8(b) are two examples,

$$x = 2 \quad \text{and} \quad x = -3$$

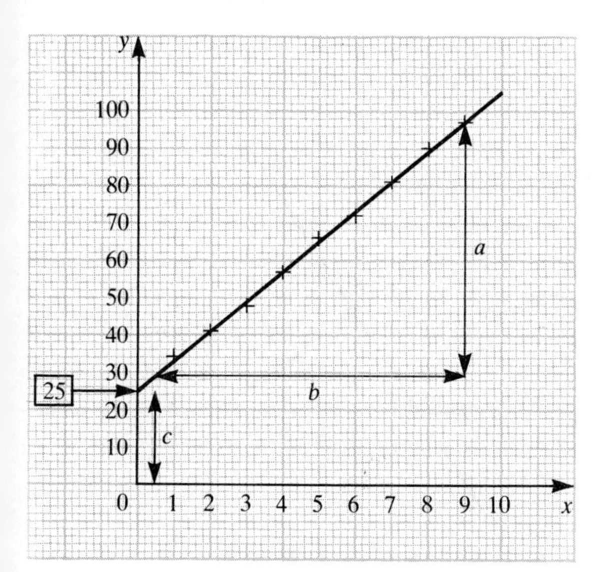

Figure 11.9 Graph for example 2.

Examples

1 A straight line passes through the points (3, 4) and (5, 12). Determine its gradient and equation.

Solution

$$\text{gradient } m = \frac{y_2 - y_1}{x_2 - x_1} = \frac{12-4}{5-3} = \frac{8}{2} = 4 \quad \textit{Ans}$$

Substituting m into the general straight-line equation, $y = mx + c$ we obtain,

$$y = 4x + c$$

and using one of the points, e.g. (3, 4) we have when $x = 3$, $y = 4 = 4 \times 3 + c$ so

$$4 = 12 + c, \quad c = 4 - 12 = -8$$

and the equation: $y = 4x - 8$ *Ans*

2 The following results were obtained in an experiment:

x values	1	2	3	4	5	6	7	8	9
y values	34	41	48	57	66	72	81	90	97

It is known that y values are likely to be accurate to better than ± 2, whilst x-values can be set to better than ± 0.01.

Plot the graph of y versus x and draw in the 'best-fit' straight line. Determine the gradient and intercept of this line and hence its equation.

Solution

The points are plotted in Figure 11.9. Due to experimental error the points do not all lie exactly on a straight line, although it is possible with a little trial and error to draw in a line which is a good comprise and 'best-fits' the data.

The gradient of the line,

$$m = \frac{a}{b} = \frac{97-30}{9-0.5} = \frac{67}{8.5} = 7.9 \quad \textit{Ans}$$

and its intercept on the y-axis, $c = 25$ *Ans.*
Hence substituting $m = 7.9$ and $c = 25$ into the general $y = mx + c$ equation, we have for our line:

$$y = 7.9x + 25 \quad \textit{Ans.}$$

3 (a) Show that the equation of a straight line expressed in the form:

$$Ax + By = C$$

has a gradient $m = \dfrac{-A}{B}$

and an intercept on the y-axis of $c = \dfrac{C}{B}$

(b) Find the gradients and y-axis intercepts of the lines:

$$5x + 7y = 35; \quad 3x - 2y = 4$$

(c) Plot the lines defined in (b) on the same $y - x$ graph and hence solve the simultaneous equations:

$$5x + 7y = 35 \quad \text{and} \quad 3x - 2y = 4$$

Solution

(a) Make y the subject of the equation, i.e.

$$Ax + By = C$$
$$By = -Ax + C$$
$$y = -\frac{A}{B}x + \frac{C}{B}$$

This is now in the standard straight-line equation form of $y = mx + c$, so on comparing like terms:

$$\text{gradient } m = -\frac{A}{B}$$

$$y\text{-axis intercept } c = \frac{C}{B} \quad \textit{Ans}$$

(b) Converting $5x + 7y = 35$ into the standard straight-line equation form, we have

$$y = -\frac{5}{7}x + 5 \qquad (1)$$

so its gradient $m = -\frac{5}{7}$ and y-axis intercept $c = 5$ *Ans*

Likewise $3x - 2y = 4$ can be expressed as

$$y = \frac{3}{2}x - 2 \qquad (2)$$

so in this case $m = \frac{3}{2}$, $c = -2$ *Ans*

(c) Equations (1) and (2) which are identical to the simultaneous equations to be solved are plotted in Figure 11.10. The coordinates of the point of intersection of the two lines gives the solution of the equations, i.e. the values of x and y that simultaneously satisfy both equations. Reading-off the coordinates, point P in Figure 11.10, we obtain:

$$x = 3.2; \quad y = 2.7 \quad Ans$$

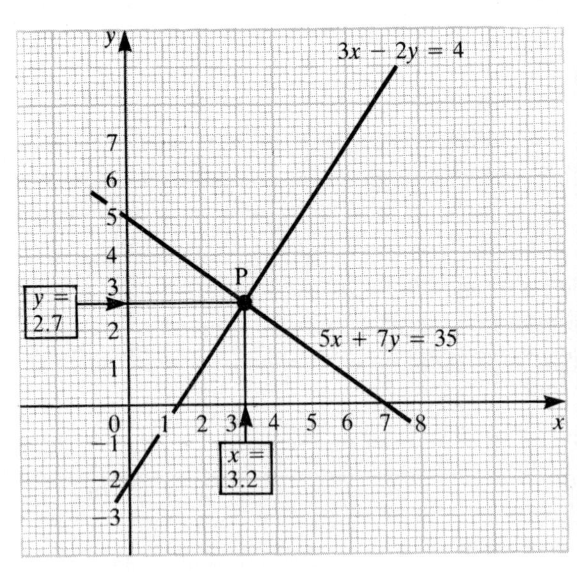

Figure 11.10 Plot of equations $5x + 7y = 35$ and $3x - 2y = 4$ and their graphical solution. See example 3(a).

Test and problems 11

Multiple choice test: MT 11

Answer block:

Question No.	0	1	2	3	4	5	6	7	8
Answer	c								

Enter your answer, that is a, b, c or d in the column under the question number in the answer block above. Note that question Qu. 0 has already been worked out and the answer inserted.

Qu. 0 Determine the intercept on the y-axis of the straight line $y = 5x - 6$
Ans (a) 5 (b) 7 (c) -6 (d) 6

Solution

The intercept the line makes with the y-axis corresponds to the value of y when $x = 0$, so when $x = 0$

$$y = (5 \times 0) - 6 = 0 - 6 = -6 \quad Ans$$

Hence the correct answer is (c) and we therefore insert 'c' under Qu. 0 in the answer block as shown.

Now carry on with the test.

Qu. 1 Which points do *not* lie on the straight line defined by the equation $y = 4x$?
Ans (a) (0, 0) (b) (1, 4) (c) $(-2, 8)$ (d) (3, 12)

Qu. 2 Determine the gradient of the line joining the points $(-1, 1)$ and $(3, 9)$.
Ans (a) 3 (b) 2 (c) 5 (d) 4

Qu. 3 Determine the equation of the straight line passing through the two points (0, 5) and (5, 5).
Ans (a) $y = 5$ (b) $y = 5x + 5$ (c) $x = 5$ (d) $y = x + 5$

Qu. 4 Determine the equation of the straight line which has a gradient of 3 and an intercept on the y-axis of 8.
Ans (a) $y = 3x - 8$ (b) $y = 3x + 8$ (c) $3y = x + 8$ (d) $y = 8x + 3$

Qu. 5 Determine graphically or otherwise the point of intersection of the straight lines: $y=x$ and $y=3x+4$

Ans (a) (0, 4) (b) (2, 2) (c) (4, −2) (d) (−2, −2)

Qu. 6 The following results were obtained in an experiment to test the current–voltage relationship of a resistor:

Current I	0	0.5	1.0	1.5	2.0	2.5
Voltage V	0	19.9	40.0	60.1	80	100

Which of the following equations best describes the $I-V$ relationship?

Ans (a) $V=50I$ (b) $I=40V$ (c) $V=40I$ (d) $V=39I$

Qu. 7 State which of the following straight lines has a zero gradient.

Ans (a) $y=5x$ (b) $y=10$ (c) $x=0$ (d) $y=-x$

Qu. 8 Solve graphically or otherwise the simultaneous equations:

$$x+y=4;\quad 3x-4y=5$$

Ans (a) $x=1$, $y=3$ (b) $x=3$, $y=1$ (c) $x=2$, $y=2$ (d) $x=6$, $y=-2$

Problems 11

1 Determine the intercept on the y-axis and the gradient of the straight lines:
(a) $y=3x+2$ (b) $y=x$ (c) $y=2x-3$ (d) $y=6$ (e) $y=-4x+7$ (d) $x=10$

2 Determine the gradient of the straight line passing through the following pairs of points:
(a) (0, 0) and (6, 3) (b) (1, 3) and (4, 7) (c) (−2, −3) and (2, 3) (d) (2, 10) and (5, 4)

3 Determine the equation of the straight lines:
(a) gradient 3 and intercept on y-axis of 12
(b) gradient of −3 and intercept on y-axis of −12
(c) line parallel to x-axis with intercept on y-axis of −3
(d) line parallel to y-axis passing through the point (5, 0)

4 Determine the equation of the straight lines passing through the points:
(a) (0, 0), (3, 3), (10, 10)
(b) (3, 3) and (6, 12)
(c) (−1, 4) and (7, −20)

5 Using the fact that 1 litre = 1.76 pints construct a graph with litres as the independent variable (x value) and pints as the dependent variable (y-value). Use your graph to convert
(a) 27 litres to pints; (b) 4.5 litres to pints; (c) 12 pints to litres.

6 Degree Celsius (°C) can be converted to degrees Fahrenheit (°F) using the formula

$$f=\frac{9}{5}c+32$$

where f = temperature in °F and c = temperature in °C.

Plot a graph of f versus c over the range $c=0$ to 100. Use the graph to convert (a) 60°C to °F, (b) 20°C to °F (c) 180°F to °C.

7 Plot the graph of $y=4x-3$ over the range $x=-5$ to $+5$. Use the graph to determine
(a) the value of y when $x=-2.8$
(b) the value of x when $y=5.6$
(c) the gradient of the straight line
(d) to solve the equations, $y=4x-3$ and $y=2x+1$

8 The following results were obtained when a length $L=1$m of mild steel wire of radius $r=0.564$ mm was subjected to tensile stress.

Tensile force F newtons	0	40	80	120	160	200	240
Extension x mm	0	0.19	0.38	0.57	0.76	1.04	1.42

Given that

$$\text{stress}=\frac{\text{force}}{\text{area}}=\frac{F}{\pi r^2},\quad \pi=3.142$$

$$\text{strain}=\frac{\text{extension in metres}}{\text{original length}}=\frac{x\text{ (in metres)}}{L}$$

plot a graph of stress (y-axis) versus strain and determine over the linear region the gradient. (The gradient is in fact Young's modulus of elasticity for the wire material.)

Part Three: Geometry and Trigonometry

12 Calculation of areas and volumes

General learning objectives: to calculate areas and volumes of plane figures and common solids using given formulae.

12.1 Areas of triangle, square, rectangle, parallelogram, circle and semi-circle

An area is a measure of this surface covered by a figure or enclosed by a boundary. Areas are quantified by multiplying two lengths and are therefore measured in square units of length. In the metric system the basic unit of length is the metre (m), so area is measured in square metres (m^2) Metric sub-multiple and multiples units of area are:

1 square millimetre,
$1\,mm^2 = 10^{-3} \times 10^{-3} = 10^{-6}\,m^2$
1 square centimetre,
$1\,cm^2 = 10^{-2} \times 10^{-2} = 10^{-4}\,m^2$
1 are $= 100\,m^2$
1 hectare $= 100$ ares $= 10^4\,m^2$

In the Imperial system we still, of course, use square inches, square yards, acres, etc. Imperial to metric conversion factors are listed below:

1 square inch $= 6.4516\,cm^2$ or $645.16\,mm^2$
1 square foot $= 929.03\,cm^2$
1 square yard $= 0.836\,127\,m^2$
1 acre $= 4840$ square yards
$= 4046.86\,m^2 = 0.40468$ hectares
1 square mile $= 640$ acres $= 259.0$ hectares

12.1.1 Area of triangle

The area of a triangle is given by the formula

$$\text{Area} = \frac{1}{2} \times \text{base} \times \text{perpendicular height}$$

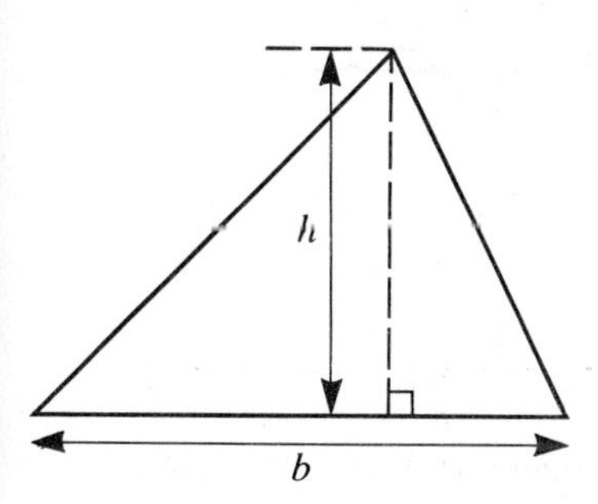

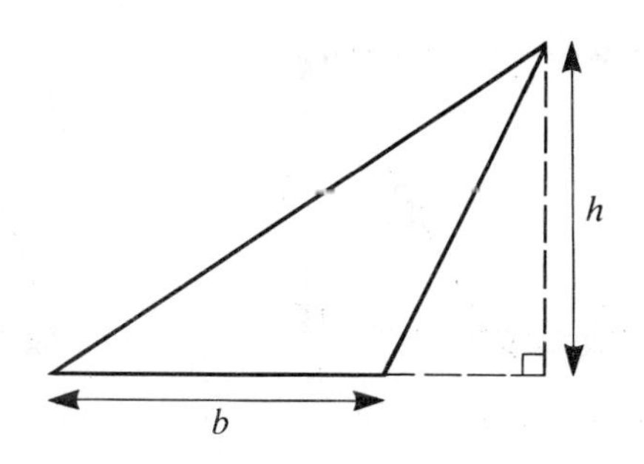

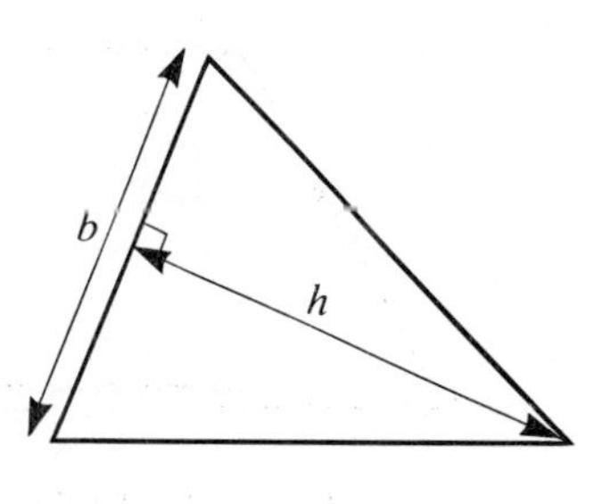

Figure 12.1 Area of triangle, $A = \frac{1}{2}bh$.

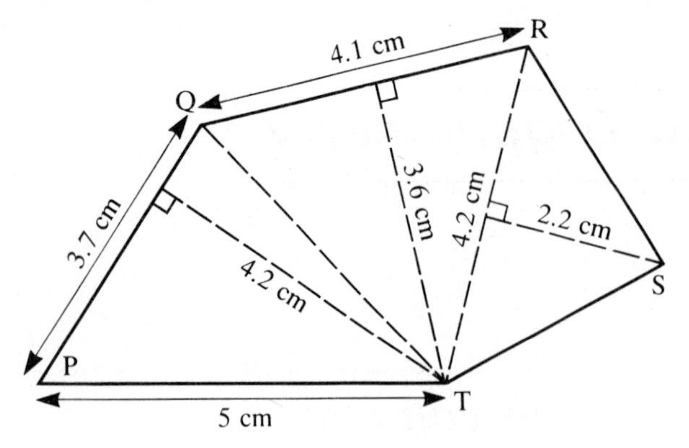

Figure 12.2 Triangulation of a plane figure to determine its area.

so for the triangles drawn in Figure 12.1 we have

$$A = \frac{1}{2}bh$$

Note we can select any one of the three sides as base b, but once this is chosen the perpendicular height h must be drawn from the angle opposite b to meet b, or b extended at right angles.

Areas of plane figures bounded by straight lines may be calculated by dividing the figure into triangles. For example the area of the 5-sided figure (a pentagon) shown in Figure 12.2 is,

$$\begin{aligned}
&\text{Area PQRST} \\
&= \text{areas of } \triangle\,\text{PQT} + \triangle\,\text{TQR} + \triangle\,\text{RST} \\
&= \left(\frac{1}{2} \times 3.7 \times 4.2\right) + \left(\frac{1}{2} \times 4.1 \times 3.6\right) \\
&\quad + \left(\frac{1}{2} \times 4.2 \times 2.2\right) \\
&= 7.77 + 7.38 + 4.62 = 19.77\ \text{cm}^2
\end{aligned}$$

12.1.2 Areas of quadrilaterals: squares, rectangles, parallelograms and trapeziums

A quadrilateral is a plane figure bounded by four straight lines, see Figure 12.3(a).

A special case of a quadrilateral is a square, in which all sides are equal and the four internal angles are right angles (90°), see Figure 12.3(b). The area of a square is

$$A = x \times x = x^2 \quad \text{where } x = \text{length of one side.}$$

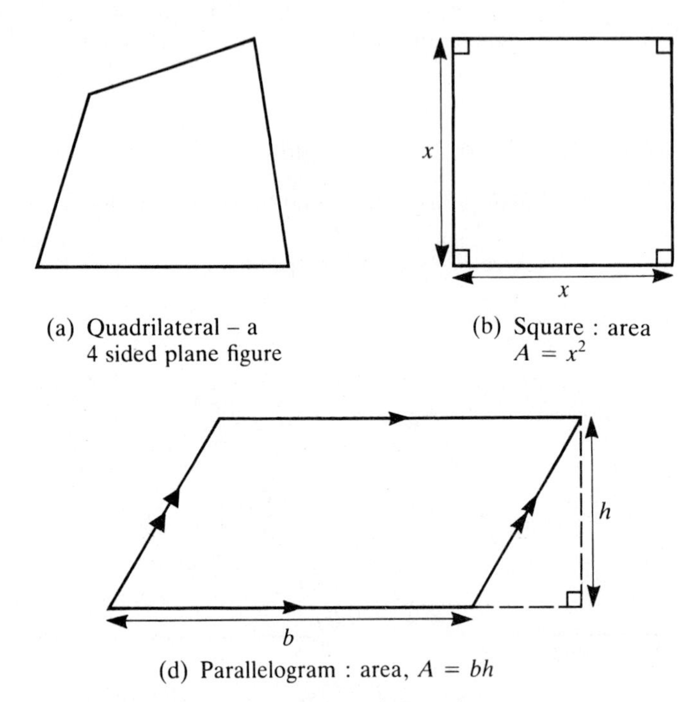

(a) Quadrilateral – a 4 sided plane figure

(b) Square : area $A = x^2$

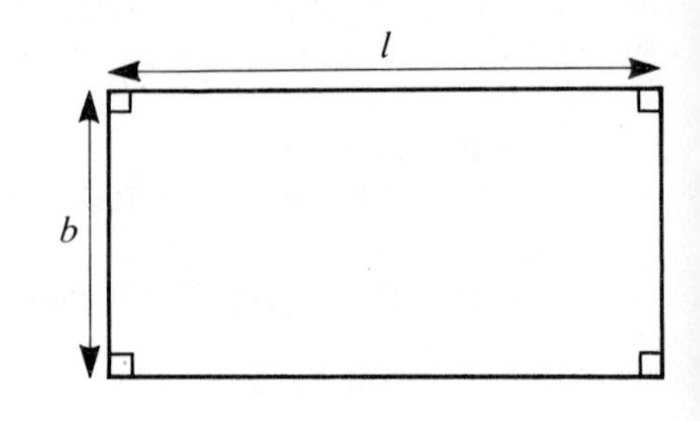

(c) Rectangle : area $A = bl$

(d) Parallelogram : area, $A = bh$

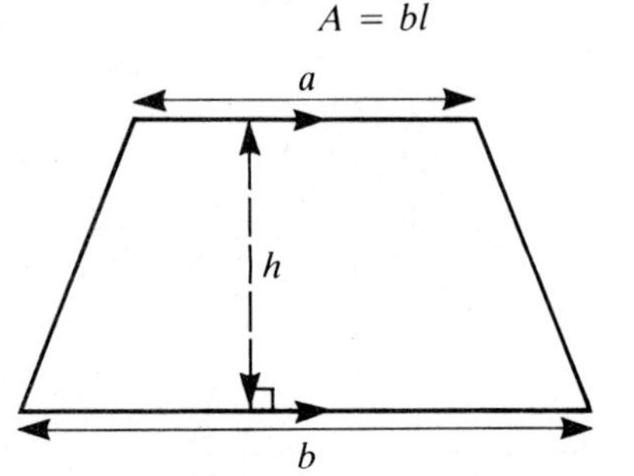

(e) Trapezium : area, $A = \frac{1}{2}(a + b)h$

Figure 12.3 Areas of common quadrilaterals.

Another special case is the rectangle, where all four internal angles are right angles but the adjacent sides differ in length, see Figure 12.3(c). The area of a rectangle is

$$A = l \times b = lb, \quad \text{where } l, b = \text{lengths of sides.}$$

A parallelogram is a quadrilateral in which the opposite sides are parallel and also equal, see Figure 12.3(d). The area of a parallelogram is

$$A = b \times h = bh,$$

where b = length of one side and h = perpendicular distance between the two sides of length b.

A special case of the parallelogram is the rhombus in which all four sides are equal.

A trapezium is a quadrilateral with one pair of sides parallel, see Figure 12.3(e). The area of a trapezium is

$$A = \frac{1}{2} \times \text{sum of parallel sides} \times \text{perpendicular distance between them}$$

$$= \frac{1}{2}(a + b)h$$

12.1.3 Areas of circles and semi-circles

The area of a circle, see Figure 12.4(a), is expressed in terms of a constant π and its radius, by

$$A = \pi r^2$$

where r = radius

$$\pi = \frac{\text{circumference}}{\text{diameter}} = 3.14159 \text{ (to 5 decimal places).}$$

The circumference of a circle is the total perimeter of the circle, i.e. the distance around the circle boundary. The diameter is the distance across the circle passing through its centre,

$$\text{diameter } d = 2r$$

so in terms of the diameter, the area of a circle is

$$A = \pi r^2 = \pi(d/2)^2 = \frac{\pi d^2}{4}$$

The area of a semi-circle, half a circle, see Figure 12.4(b) is

$$A_{sc} = \frac{1}{2}\pi r^2 \quad \text{or} \quad \frac{1}{8}\pi d^2$$

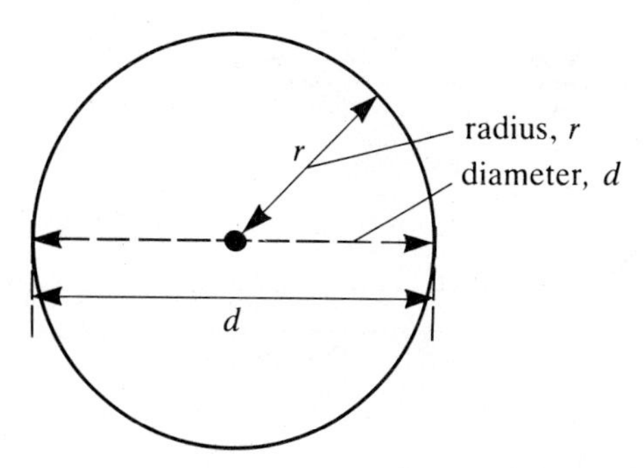

(a) Area of circle, $A = \pi r^2 = \frac{1}{4}\pi d^2$

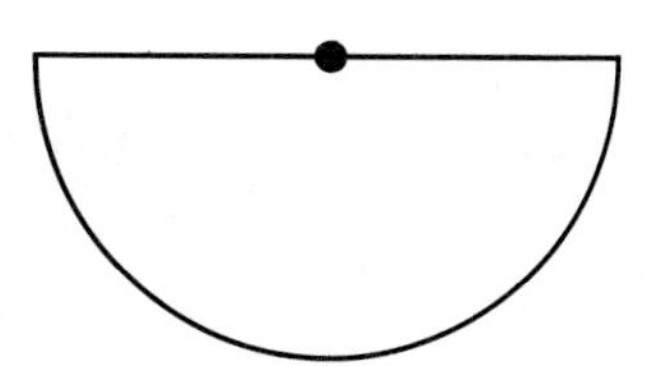

(b) Area of semi-circle, $A_{SC} = \frac{1}{2}\pi r^2 = \frac{1}{8}\pi d^2$

Figure 12.4 Areas or circle and semi-circle.

Examples

1 A building plot consists of a rectangle of dimensions 14.5 m by 37.8 m. Calculate its area in square metres and also in acres and hectares, given 1 acre = 4046.86 m^2.

Solution

Area $A = lb = 37.8 \times 14.5 = 548.1 \text{ m}^2$ *Ans*

Area in acres $= \frac{548.1}{4046.86} = 0.1354$ acres *Ans*

1 hectare $= 10^4 \text{ m}^2$,
so area = 0.05481 hectares *Ans*

2 Calculate the area of the figure shown in Figure 12.5 which consists of semi-circle plus rectangle plus triangle (or semi-circle plus a trapezium). Take $\pi = 3.142$

Solution

Area of semi-circle, diameter $d = 6$ cm

$$A_1 = \frac{1}{8}\pi d^2 = \frac{1}{8} \times 3.142 \times 6^2 = 14.14 \text{ cm}^2$$

Area of rectangle, XYVW,

$$A_2 = 9.2 \times 6 = 55.2 \text{ cm}^2$$

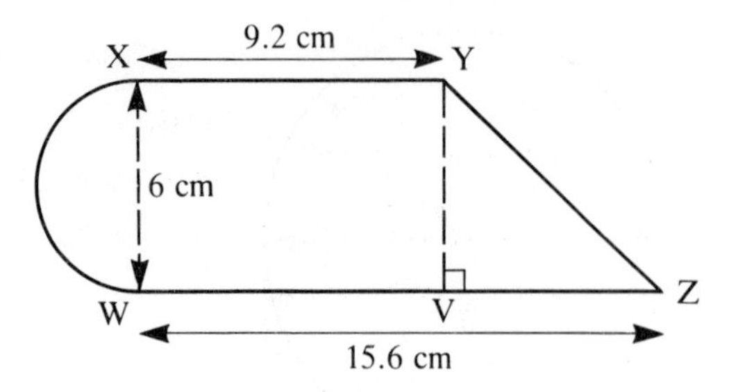

Figure 12.5 For example 2.

Area of triangle, YVZ

$$A_3 = \frac{1}{2}bh$$

where $h = 6$ cm and $b = \text{VZ} = 15.6 - 9.2 = 6.4$ cm

so $A_3 = \frac{1}{2} \times 6.4 \times 6 = 19.2\,\text{cm}^2$

Hence total area,

$$A = A_1 + A_2 + A_3 = 14.14 + 55.2 + 19.2$$
$$= 88.54\,\text{cm}^2 \quad \textit{Ans}$$

3 Calculate the shaded area shown in Figure 12.6, given that the diameter D of the larger outer circle is 40 mm and the diameter d of each of the three smaller circles is 15 mm. Take $\pi = 3.142$

Solution

Area of larger circle, $A = \frac{1}{4}\pi D^2$

$$= \frac{1}{4} \times 3.142 \times 40^2$$
$$= 1256.8\,\text{mm}^2$$

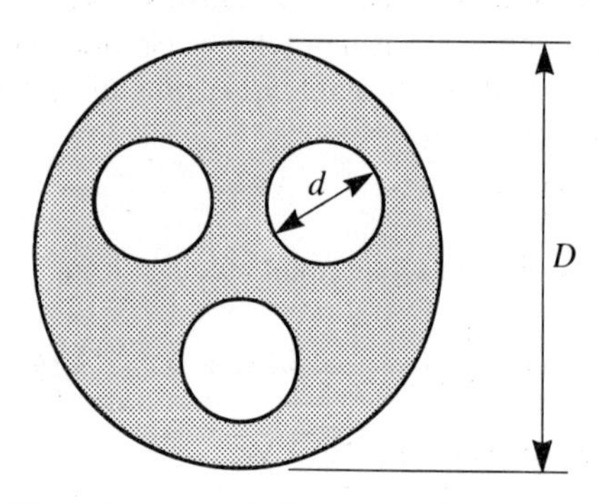

Figure 12.6 For example 3.

Area of smaller circle, $a = \frac{1}{4}\pi d^2$

$$= \frac{1}{4} \times 3.142 \times 15^2$$
$$= 176.7\,\text{mm}^2$$

So area of shaded region $= A - 3a$

$$= 1256.8 - (3 \times 176.7)$$
$$= 726.7\,\text{mm}^2 \quad \textit{Ans}$$

12.2 Volumes of common solids: cubes, cylinders, prisms

Volume is a measure of the bulk or capacity of a body. The magnitude of a volume depends on the product of three lengths and is therefore measured in cubic units. In the metric system, since the basic unit of length is the metre (m), the unit of volume is the cubic metre (m^3). Metric and some common Imperial units used to quantify volume are given below:

$1\,\text{m}^3 = 100 \times 100 \times 100\,\text{cm}^3 = 10^6\,\text{cm}^3$
$1\,\text{litre} = 1000\,\text{cm}^3 = 10^{-3}\,\text{m}^3$
Imperial units and metric equivalents:
1 cubic inch $= 16.387\,\text{cm}^3$
1 cubic foot $= 1728$ cubic inch $= 0.028\,316\,8\,\text{m}^3$
1 cubic yard $= 27$ cubic feet $= 0.764\,55\,\text{m}^3$
1 gallon $= 8$ pints $= 4.546\,092$ litres
$= 4.546 \times 10^{-3}\,\text{m}^3$

12.2.1 Volumes of cubes

A cube is a solid contained by six equal squares, see Figure 12.7(a). The volume of a cube is equal to the product of its three equal lengths, i.e. the cube of its face length:

$$v = x^3, \quad \text{where } x = \text{length of side of square face.}$$

For solid figures bounded by rectangles, the volume is given by the product of its external dimensions, i.e. with reference to Figure 12.7(b),

$$v = h \times b \times l = hbl$$

12.2.2 Volume of cylinders and spheres

The volume of a cylinder, see Figure 12.8, is given by

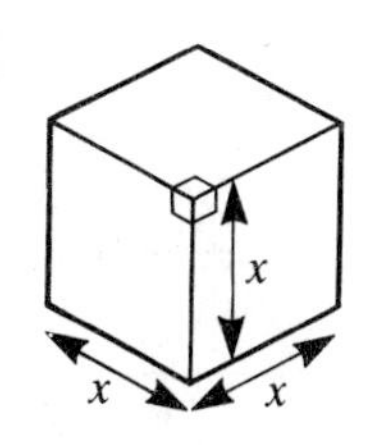

(a) Cube : volume, $v = x^3$

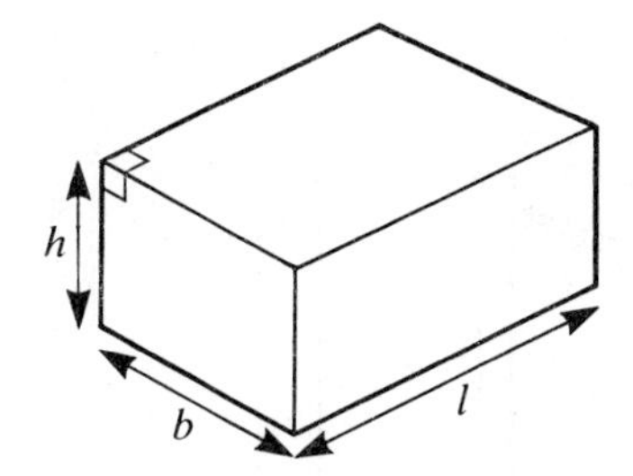

(b) Cuboid: volume $v = hbl$

Figure 12.7.

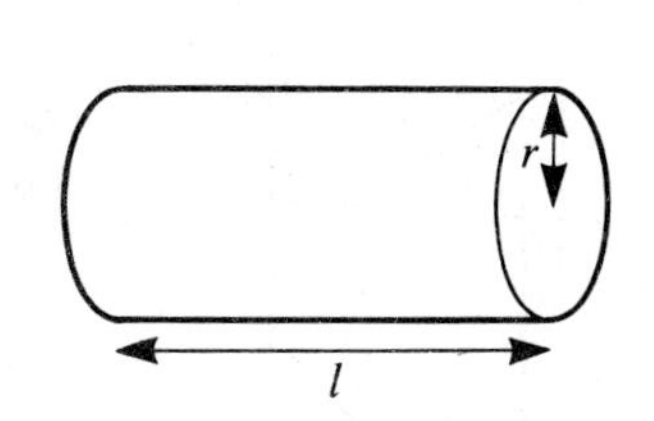

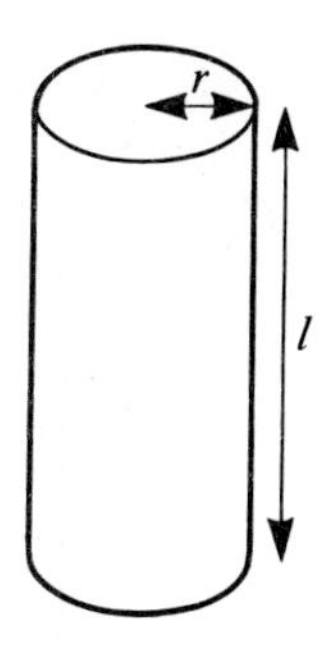

Volume of cylinder, $v = \pi r^2 l$

Figure 12.8.

$$v = \text{area of circular cross-section} \times \text{length or height}$$
$$= \pi r^2 \times l = \pi r^2 l$$

where r = radius of cylinder's circular cross-section
l = length or height of cylinder
π = 3.14159 (to 5 decimal places)

The volume of a sphere, see Figure 12.9, is

$$v = \frac{4}{3}\pi r^3$$

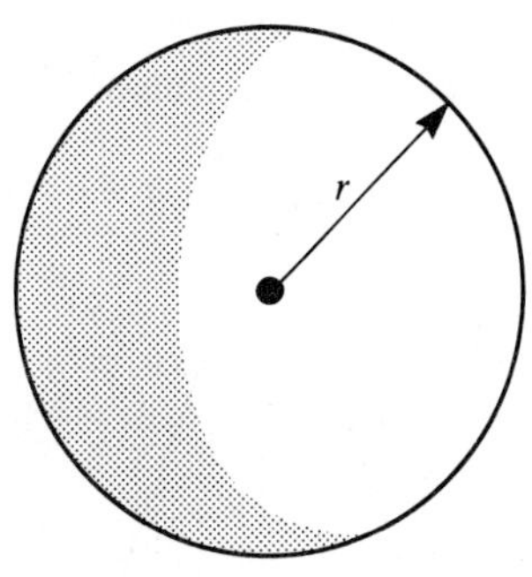

Volume of sphere, $v = \frac{4}{3}\pi r^3$

Figure 12.9.

12.2.3 Volumes of prisms

A prism is a solid figure which has a constant cross-sectional area and shape throughout its length. Cubes and cylinders are particular forms of prisms with, respectively, square and circular cross-sectional shapes.

The formula for the volume of a prism is

$$\text{volume} = \text{cross-sectional area} \times \text{length}$$

so for the prisms shown in Figure 12.10 we have,

(a) rectangular prism, $v = (a \times b) \times l = abl$

(b) triangular prism, $v = \left(\frac{1}{2}b \times h\right) \times l = \frac{1}{2}bhl$

(c) trapezoidal prism, $v = \left[\frac{1}{2}(a + b) \times h\right] \times l$

$$= \frac{1}{2}hl(a + b)$$

(d) hexagonal prism, $v = A \times l$

where A = cross-sectional area of hexagon
= area of 6 identical triangles, when hexagon is 'regular' as shown in Figure 12.10(d)

$$= 6 \times \frac{1}{2}bh = 3bh$$

12.2.4 Volumes of cones and pyramids

The volume of a cone, see Figure 12.11, is given by

$$v = \frac{1}{3} \times \text{base area} \times \text{perpendicular height}$$
$$= \frac{1}{3}\pi r^2 h$$

where r = radius of circular base
h = perpendicular height of cone

The volume of a pyramid with a rectangular base as shown in Figure 12.12, is given by

$$v = \frac{1}{3} \times \text{base area} \times \text{perpendicular height}$$
$$= \frac{1}{3}abh$$

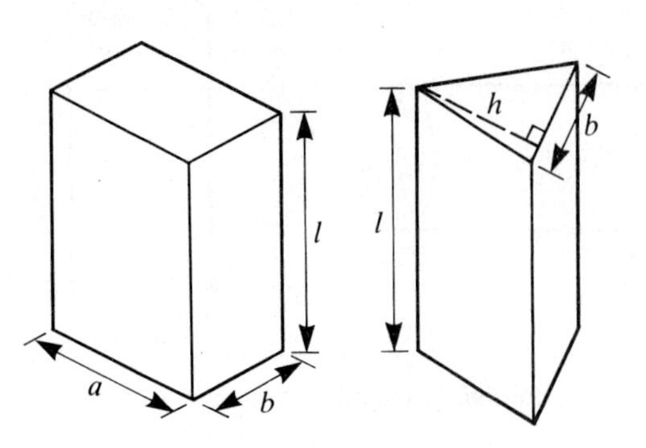

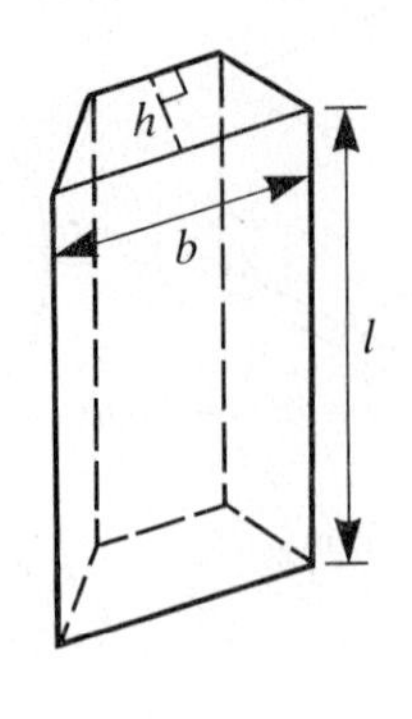

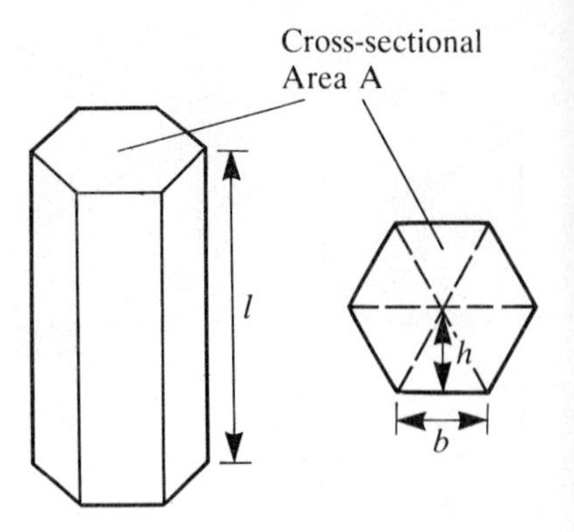

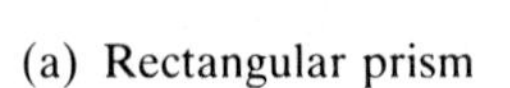
(a) Rectangular prism

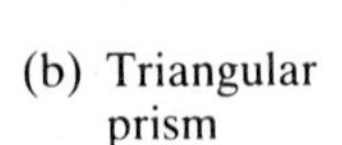
(b) Triangular prism

(c) Trapezoidal prism

(d) Hexagonal prism

Figure 12.10 Volume of prisms: volume $v = Al$, where A = cross-sectional area l = length of prism.

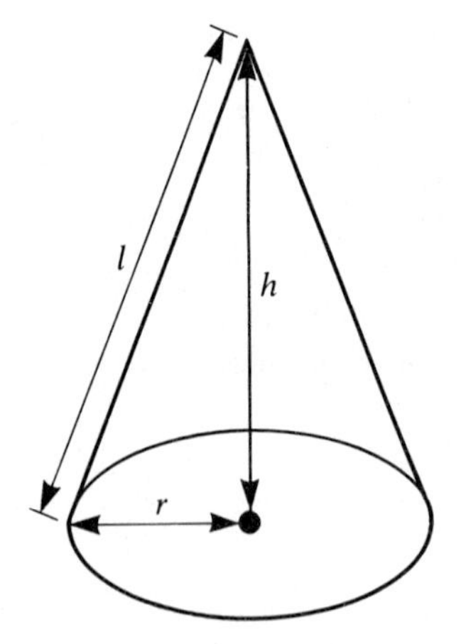

Figure 12.11 Volume of cone, $v = \frac{1}{3}\pi r^2 h$ (l = slant length).

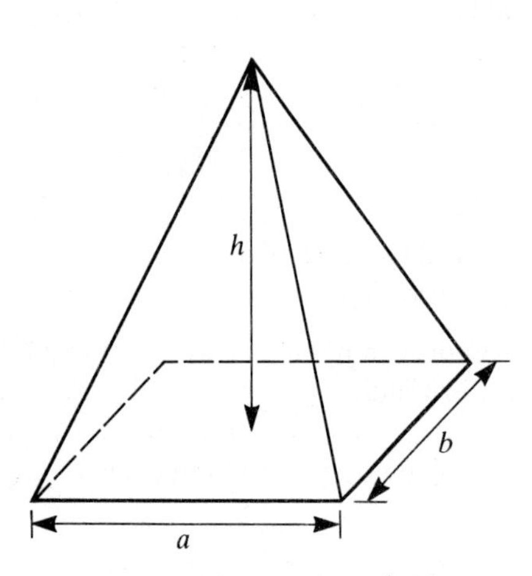

Figure 12.12 Volume of pyramid, $v = \frac{1}{3}abh$.

Examples

1 Calculate the volume of a tank of rectangular cross-section 0.75 m by 1.2 m and of height 1.5 m. Given that 1 litre $= 10^{-3}\,\text{m}^3$ and 1 gallon = 4.546 litres, determine the capacity of the tank in litres and gallons.

Solution

Volume of tank

$v = 0.75 \times 1.2 \times 1.5 = 1.35\,\text{m}^3$ *Ans*

$= 1.35 \div 10^{-3} = 1350$ litres *Ans*

$= 1350 \div 4.546 = 297.0$ gallons *Ans*

2 Calculate the volumes of the solids shown in Figure 12.13. Take $\pi = 3.142$.

Solution

(a) Volume of cylinder shown in Figure 12.13(a),

$$v = \pi r^2 l$$
$$= 3.142 \times 9^2 \times 10 = 2545.0\,\text{mm}^3 \quad \textit{Ans}$$

(b) Volume of triangular prism shown in Figure 12.13(b),

$$v = A \times l$$

where $A = \frac{1}{2}bh = \frac{1}{2} \times 1 \times 0.6 = 0.3\,\text{m}^2$, triangular cross-section

$l = 1.5$ m, height of prism

so $v = 0.3 \times 1.5 = 0.45\,\text{m}^3$ *Ans*

(c) Volume of cone shown in Figure 12.13(c)

$$v = \frac{1}{3}\pi r^2 l$$

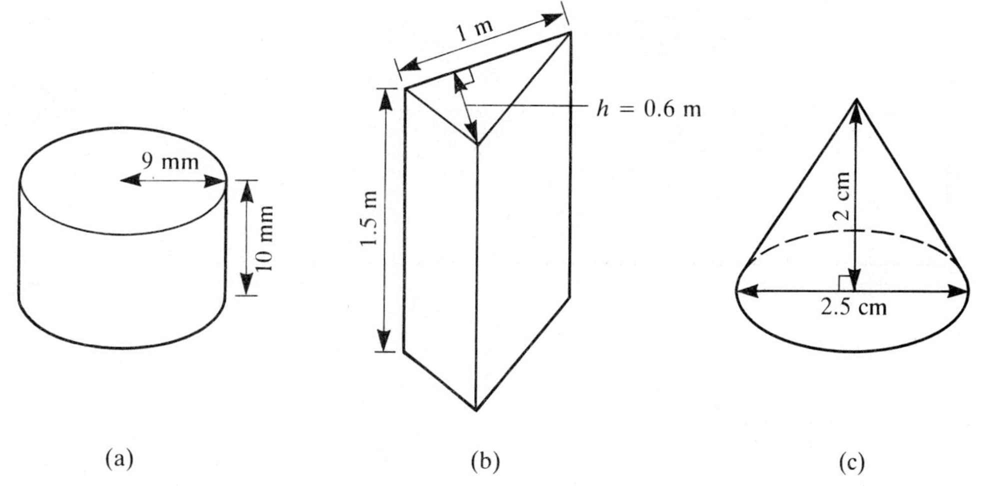

Figure 12.13 For example 2 (a) (b) (c).

$$= \frac{1}{3} \times 3.142 \times (2.5/2)^2 \times 2$$

$$= 3.273 \text{ cm}^3 \quad \textit{Ans}$$

3 The density of copper is given by $8930\,\text{kg/m}^3$, i.e. a cubic metre of copper weighs 8930 kg. Calculate the mass of
(a) a cube of copper of side 5 cm,
(b) a sphere of copper of radius 10 mm.

Solution

(a) Volume of cube $v = 0.05 \times 0.05 \times 0.05$
$= 0.05^3 = 1.25 \times 10^{-4}\,\text{m}^3$
mass of copper = density $\times\, v$
$= 8930 \times 1.25 \times 10^{-4}$
$= 1.116\,25\,\text{kg}$ *Ans*

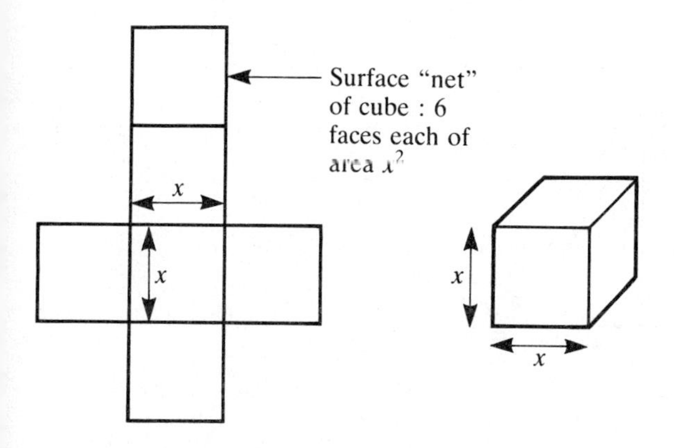

Figure 12.14 Surface area of cube, $S = 6x^2$.

(b) Volume of sphere,

$$v = \frac{4}{3}\pi r^3$$

$$= \frac{4}{3} \times 3.142 \times (10 \times 10^{-3})^3$$

$$= 4.1893 \times (10^{-2})^3 = 4.1893 \times 10^{-6}\,\text{m}^3$$

mass of copper
$= 8.930 \times 10^3 \times 4.1893 \times 10^{-6}$
$= 37.411 \times 10^{-3}\,\text{kg}$ or $37.411\,\text{g}$ *Ans*

12.3 Surface areas of cubes, prisms and cylinders

12.3.1 Surface area of a cube

The surface area of each face of a cube of side length x is x^2 so the total surface area of a cube,

$$S = 6x^2$$

as the cube has in total six faces, see Figure 12.14

12.3.2 Surface area of prisms

The surface area of a prism equals the sum of the areas of its two ends plus the area of its exterior surface along its length, i.e. with reference to Figure 12.15, the total surface area,

$$S = 2A + S_1$$

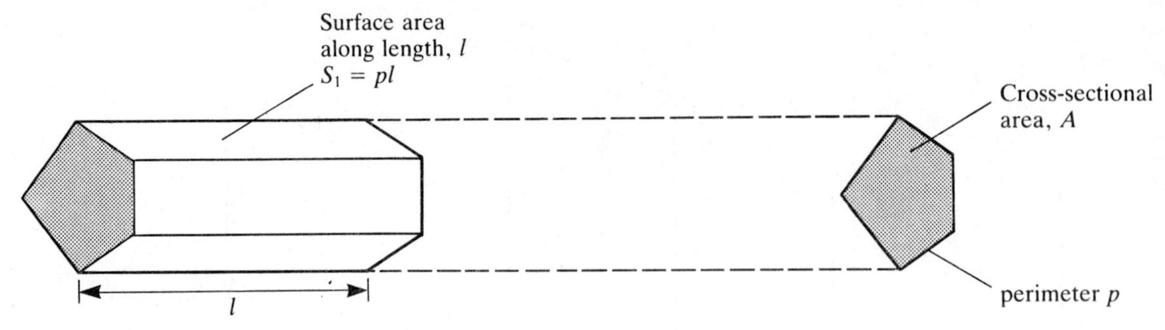

Figure 12.15 Total surface area of prism, $S = 2A + S_1$.

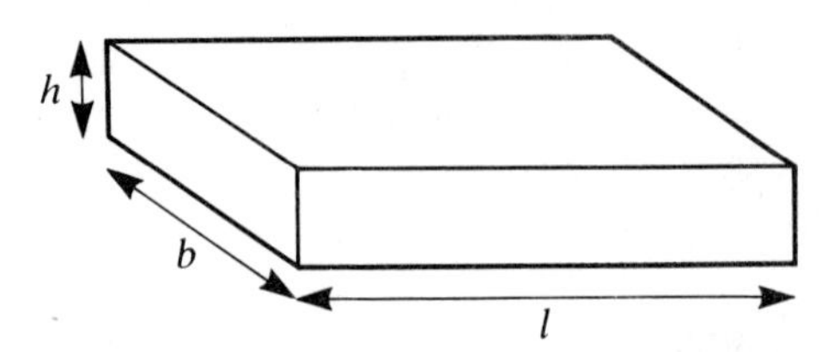

Figure 12.16 Surface area of $S = 2(hb + hl + bl)$.

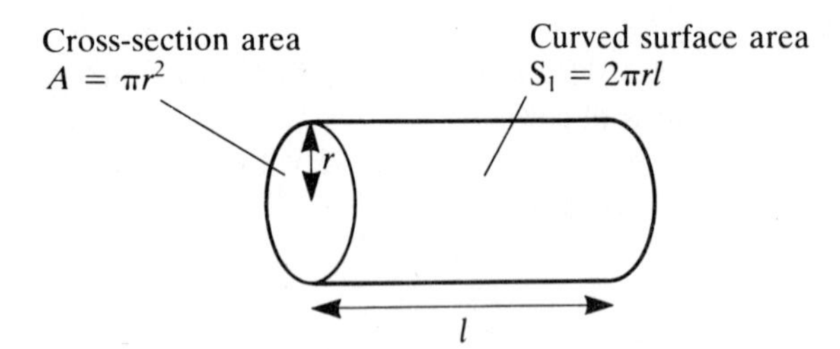

Figure 12.17 Surface area of cylinder, $S = 2\pi r^2 + 2\pi rl$.

where A = area of end, cross-sectional area
S_1 = area of surface along prism length

S_1 is found by multiplying the perimeter p of the prism's cross-section by the prism length, i.e.

$$S_1 = pl$$

Thus, for example, the total surface area for the rectangular cross-section prism of Figure 12.16 is

$$\begin{aligned} S &= 2 \times hb + (h + b + h + b)l \\ &= 2(hb + hl + bl) \end{aligned}$$

which checks, as of course it must, with sum of the areas of all six rectangular faces.

12.3.3 Surface area of cylinders and spheres

The surface area of a cylinder equals the sum of the areas of its circular ends and the area of its curved

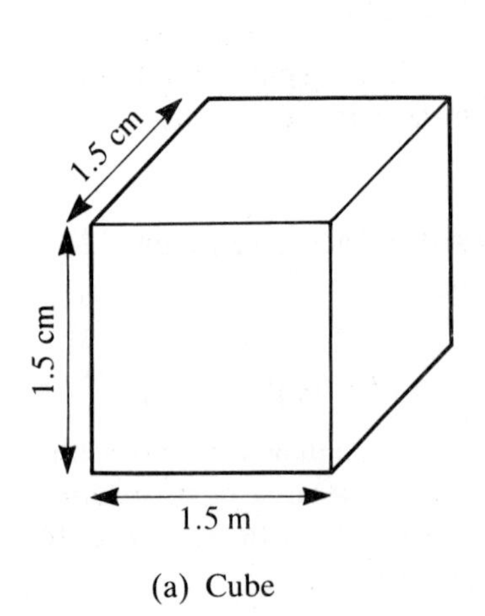

(a) Cube

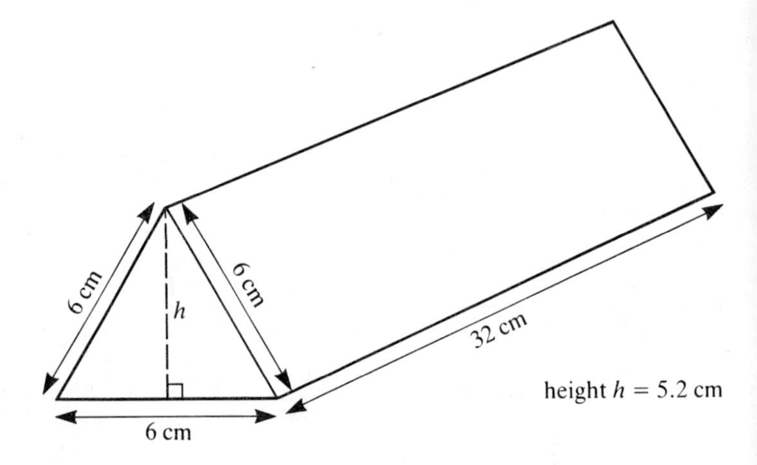

(b) Triangular prism

Figure 12.18 For example 1.

surface along its length, see Figure 12.17. The surface area is therefore given by

$$S = 2\pi r^2 + 2\pi rl$$

where πr^2 = area of each end
$2\pi r$ = circumference (perimeter) of circular cross-section
$2\pi rl$ = area of curved surface of cylinder
r = cylinder radius
l = cylinder length or height.

The cylinder is a special case of a prism with a circular cross-section.

The surface area of a sphere is given by

$$S = 4\pi r^2$$

where r = sphere radius

12.3.4 Surface area of cone and pyramid

The surface area of the 'curved' area of the cone is given by

$\pi r \times$ slant length $= \pi rl$ (see Figure 12.11)

and including the area of the circular base, the total surface area,

$$S = \pi r^2 + \pi rl$$

The total surface area of the pyramid shown in Figure 12.12,

S = base area + sum of areas of sloping triangular faces

$$= a \times b + a\sqrt{\left(h^2 + \frac{1}{4}b^2\right)} + b\sqrt{\left(h^2 + \frac{1}{4}a^2\right)}$$

Examples

1 Calculate the total surface area of the cube and triangular prism shown in Figure 12.18.

Solution

(a) Surface area of cube,

$$S = 6x^2 \quad \text{where } x = 1.5\,\text{cm}$$
$$= 6 \times 1.5^2 = 6 \times 2.25 = 13.5\,\text{cm}^2$$

(b) Surface area of triangular prism,

$S = 2 \times$ area of triangular end + area along prism length

Triangular end area $= \frac{1}{2}bh = \frac{1}{2} \times 6 \times 5.2$
$= 15.6\,\text{cm}^2$

area along prism length
$=$ perimeter $\times$ length
$= (6 + 6 + 6) \times 32 = 576\,\text{cm}^2$

so $S = (2 \times 15.6) + 576 = 607.2\,\text{cm}^2$ *Ans*

2 Calculate the total surface area and volume of the tank shown in Figure 12.19 which consists of a cylinder with a hemi-spherical top.

Solution

Surface area of hemisphere $= \frac{1}{2} \times 4\pi r^2 = 2\pi r^2$

Surface area of curved surface of cylinder $= 2\pi rl$
Surface area of cylinder base $= \pi r^2$
Total surface area
$= 2\pi r^2 + 2\pi rl + \pi r^2 = 3\pi r^2 + 2\pi rl$
$= \pi r(3r + 2l)$
$= 3.142 \times 0.35(3 \times 0.35 + 2 \times 1.5)$
$= 4.454\,\text{m}^2$ *Ans*

Volume = volume of cylinder $+ \frac{1}{2}$ volume of sphere

$$= \pi r^2 l + \frac{1}{2} \times \frac{4}{3}\pi r^3$$

$= (3.142 \times 0.35^2 \times 1.5)$
$+ (2 \times 3.142 \times 0.35^3/3)$
$= 0.577 + 0.090 = 0.667\,\text{m}^3$ *Ans*

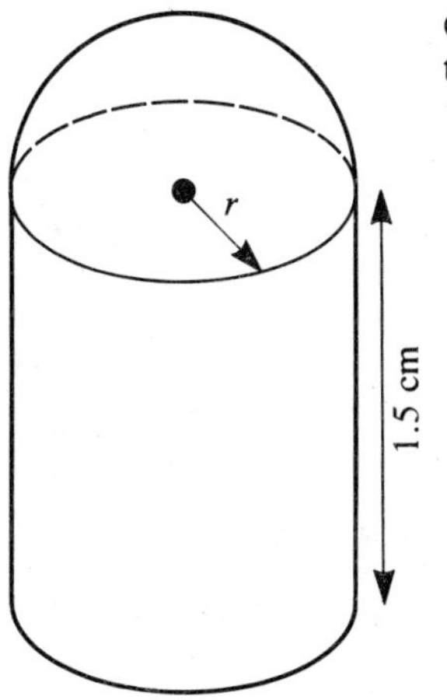

Figure 12.19 For example 2.

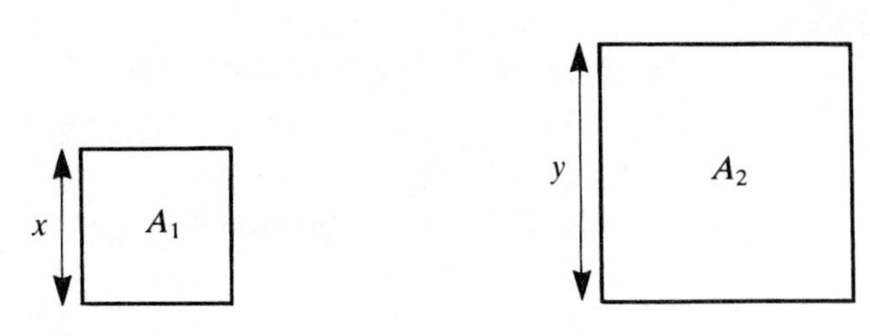

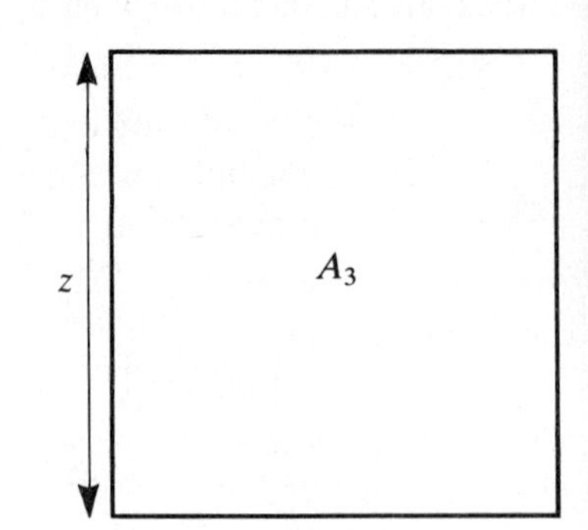

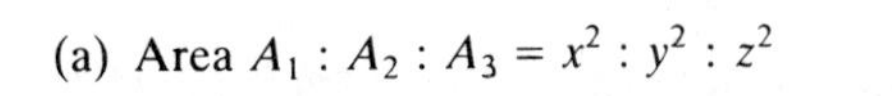

(a) Area $A_1 : A_2 : A_3 = x^2 : y^2 : z^2$

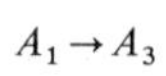

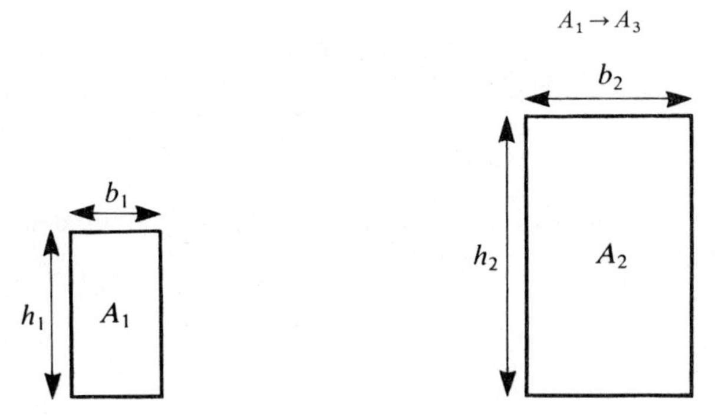

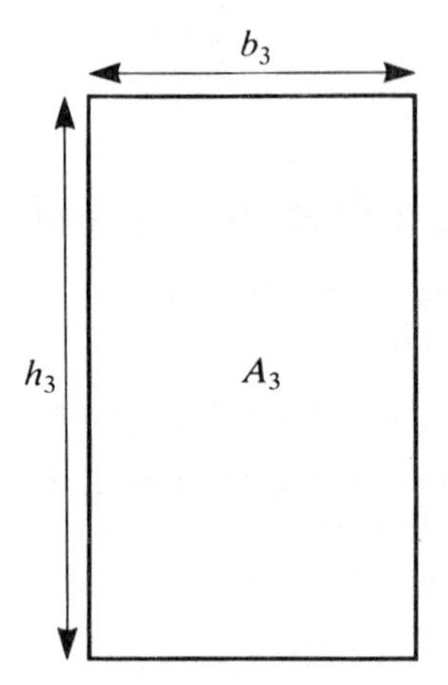

(b) Area $A_1 : A_2 : A_3 = {h_1}^2 : {h_2}^2 : {h_3}^2 = {b_1}^2 : {b_2}^2 : {b_3}^2$

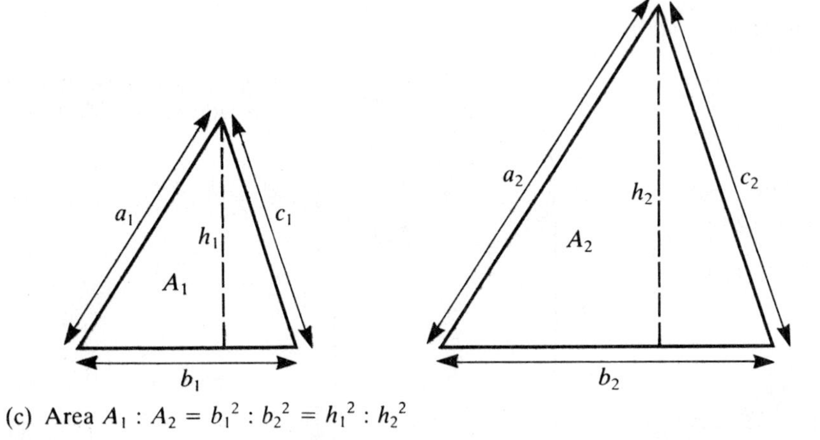

(c) Area $A_1 : A_2 = {b_1}^2 : {b_2}^2 = {h_1}^2 : {h_2}^2$

Figure 12.20 Areas of similar figures are proportional to the 'squares' of corresponding dimensions.

12.4 Proportionality for similarly shaped figures and volumes

When figures are *similar* in shape in the sense that they are exact replicas of one another either magnified or diminished, then their areas are in the direct ratio of the squares of the corresponding dimensions.

This property is illustrated in Figure 12.20. Obviously for the case of the squares where area equals the side length squared, we have:

$$A_1 = x^2, \quad A_2 = y^2, \quad A_3 = z^2$$

so $A_1 : A_2 : A_3 = x^2 : y^2 : z^2$

In Figure 12.20(b) the rectangles shown all have equal breadth to height ratios, i.e.

$$\frac{b_1}{h_1} = \frac{b_2}{h_2} = \frac{b_3}{h_3} = \text{a constant, } k\text{, say}$$

and are therefore *similar* rectangles. Their respective areas are

$$A_1 = b_1 h_1 = k h_1^2 \quad \text{or} \quad b_1^2/k$$
$$A_2 = b_2 h_2 = k h_2^2 \quad \text{or} \quad b_2^2/k$$
$$A_3 = b_3 h_3 = k h_3^2 \quad \text{or} \quad b_3^2/k$$

and so $\dfrac{A_1}{A_2} = \dfrac{k h_1^2}{k h_2^2} = \dfrac{h_1^2}{h_2^2}$

or $\dfrac{A_1}{A_2} = \dfrac{b_1^2/k}{b_2^2/k} = \dfrac{b_1^2}{b_2^2}$

and likewise $\dfrac{A_2}{A_3} = \dfrac{h_2^2}{h_3^2}$ or $\dfrac{b_2^2}{b_3^2}$

Hence the areas are in the ratio of the corresponding dimensions squared, i.e.

$$A_1 : A_2 : A_3 = h_1^2 : h_2^2 : h_3^2 \text{ or } b_1^2 : b_2^2 : b_3^2$$

For example, if the heights of the similar rectangles were respectively 3, 5, 7 then the ratio of their areas would be:

$$A_1 : A_2 : A_3 = 3^2 : 5^2 : 7^2 = 9 : 25 : 49$$

Likewise for *similar* triangles, see Figure 12.20(c), the ratio of their areas equals the ratio of the squares of corresponding dimensions:

$$\frac{A_1}{A_2} = \frac{a_1^2}{a_2^2} = \frac{b_1^2}{b_2^2} = \frac{c_1^2}{c_2^2} = \frac{h_1^2}{h_2^2}$$

So for example if two similar triangles have base length equal to 4 and 7 respectively, then the ratio of their areas:

$$A_1 : A_2 = 4^2 : 7^2 = 16 : 49$$

Note: triangles are *similar* when they have equal angles and it is a property of similar triangles that the ratios of corresponding sides are equal, i.e. with reference to Figure 12.20(c)

$$\frac{a_1}{a_2} = \frac{b_1}{b_2} = \frac{c_1}{c_2}$$

In the case of solids of similar shape in the sense that the solids are exact replicas of one another when either diminished or magnified, then their

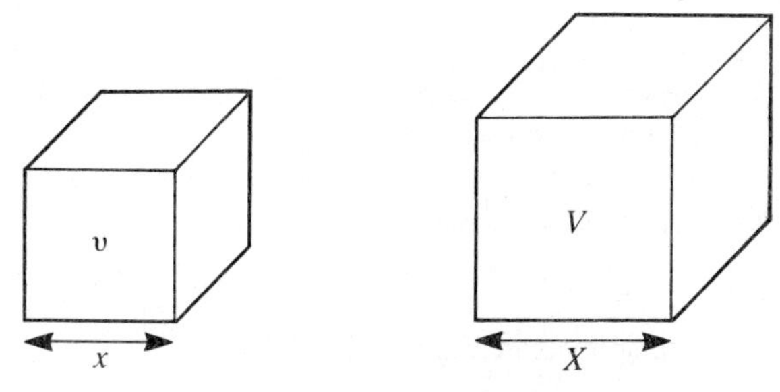

(a) Cubes : volume ratio, $v : V = x^3 : X^3$

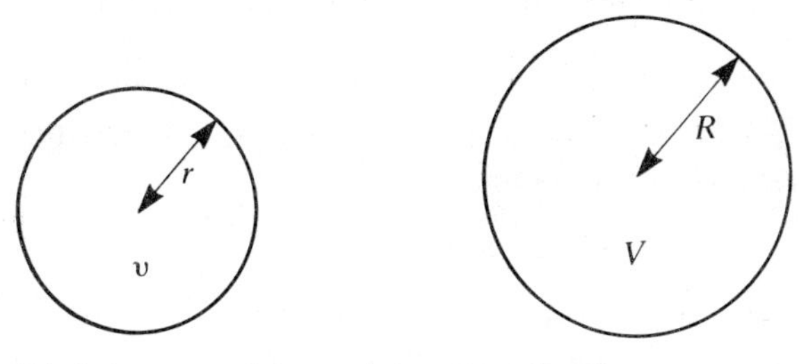

(b) Spheres : volume ratio $v : V = r^3 : R^3$

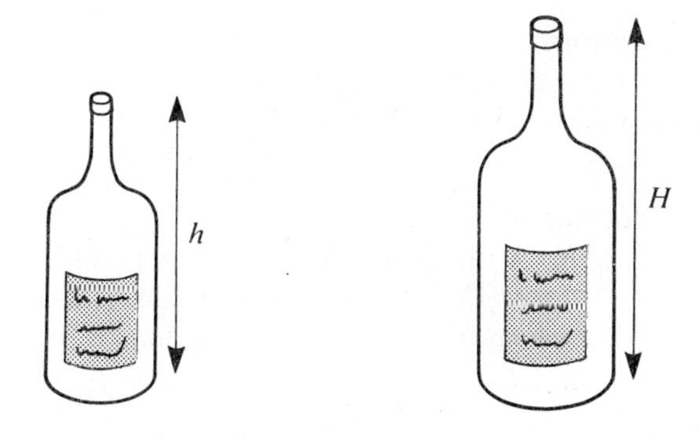

(c) Ratio of volumes of similar shaped bodies, $v : V = h^3 : H^3$

Figure 12.21 Volumes of similar solids are proportional to the 'cubes' of corresponding dimensions.

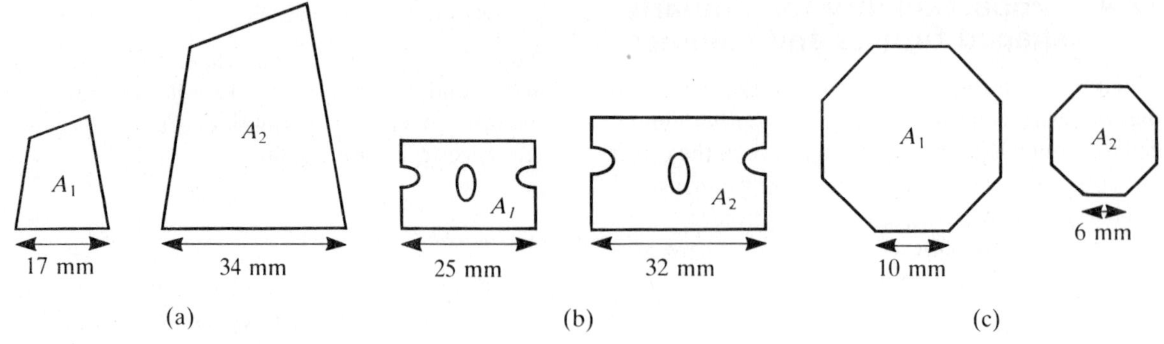

Figure 12.22 Similar figures for example 1.

volumes are in the direct ratio of the cubes of the corresponding dimensions.

This property is illustrated in Figure 12.21. Obviously for the cubes, as

$$v = x^3 \quad \text{and} \quad V = X^3, \qquad v:V = x^3:X^3$$

For the spheres in Figure 12.21(c)

$$v:V = r^3:R^3$$

whilst for the *similar* bottles of Figure 12.21(c)

$$v:V = h^3:H^3$$

So for example, if $h = 10\,\text{cm}$ and $H = 20\,\text{cm}$, the ratio of their volumes

$$v:V = 10^3:20^3 = 1^3:2^3 = 1:8$$

so a bottle of twice the height has a capacity of eight times the smaller similar one.

Examples

1 Calculate the ratio of the areas of the *similar* figures shown in Figure 12.22.

Solution

(a) Since the quadrilaterals of Figure 12.22(a) are *similar*, their areas are in direct ratio to the squares of corresponding dimensions.

Hence $A_1:A_2 = 17^2:34^2$
$= 1^2:2^2 = 1:4$ *Ans*

(b) For the *similar* 'razor blades' of Figure 12.22(b)

$$A_1:A_2 = 25^2:32^2 = 625:1024 \quad \textit{Ans}$$

or $\dfrac{A_1}{A_2} = \dfrac{625}{1024} = 0.61$ *Ans*

(c) For the similar 8-sided figures (octagons) of Figure 12.22(c)

$A_1:A_2 = 10^2:6^2 = 100:36$ or $25:9$ *Ans*

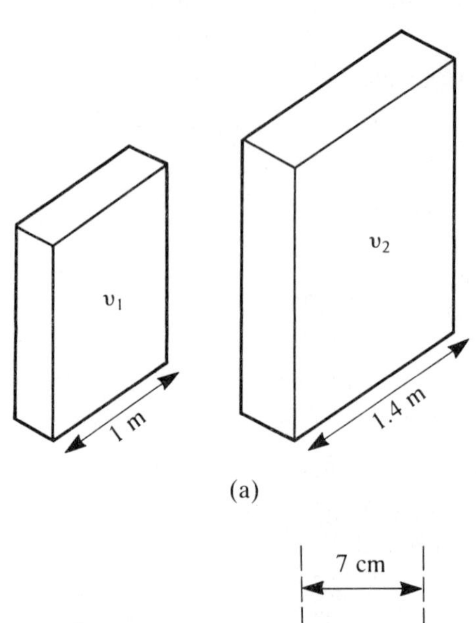

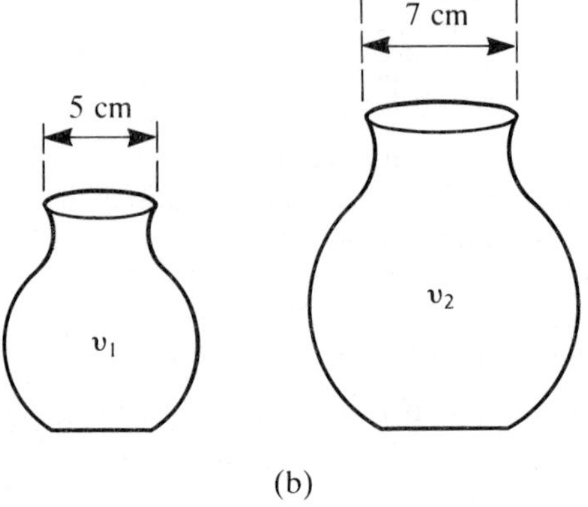

Figure 12.23 Similar solids for example 2.

2 Determine the ratios of the volumes for the similar solids shown in Figure 12.23.

Solution

Since for similar solids the ratio of volumes equals the ratio of the cubes of corresponding dimensions, we have for Figure 12.23(a),

$$v_1 : v_2 = 1^3 : 1.4^3 = 1 : 2.744 \quad \textit{Ans} \text{ (a)}$$

and for Figure 12.23(b)

$$v_1 : v_2 = 5^3 : 7^3 = 125 : 343 \quad \textit{Ans} \text{ (b)}$$

Test and problems 12

Multiple choice test: MT 12

Answer block:

Question No.	0	1	2	3	4	5	6	7	8
Answer	a								

Enter your answer, that is a, b, c or d in the column under the question number in the answer block above.
Note that question Qu. 0 has already been worked out and the answer inserted.

Qu. 0 Calculate the area of the curved surface of a cylinder of radius $r = 8$ cm and length $l = 15$ cm. Take $\pi = 3.142$

Ans (a) $754.1\,\text{cm}^2$ (b) $3275.5\,\text{cm}^2$
(c) $402.2\,\text{cm}^2$ (d) $377\,\text{cm}^2$

Solution

The curved surface area of a cylinder,

$$S_1 = 2\pi rl = 2 \times 3.142 \times 8 \times 15$$
$$= 754.1\,\text{cm}^2 \quad \textit{Ans}$$

Thus the correct answer is (a) and 'a' is entered under Qu. 0 in the answer block as shown above.

Now carry on with the test.

Qu. 1 Calculate the area of the parallelogram shown in Figure 12.24.

Ans (a) $72\,\text{cm}^2$ (b) $28.8\,\text{cm}^2$
(c) $57.6\,\text{cm}^2$ (d) $115.2\,\text{cm}^2$

Qu. 2 Calculate the area of the semi-circle shown in Figure 12.25.

Ans (a) $1256.8\,\text{m}^2$ (b) $628.4\,\text{m}^2$
(c) $157.1\,\text{m}^2$ (d) $314.2\,\text{m}^2$

Qu. 3 Calculate the volume of a cylinder of radius 1.6 m and height 2 m. Take $\pi = 3.142$.

Ans (a) $4.022\,\text{m}^3$ (b) $16.09\,\text{m}^3$
(c) $20.11\,\text{m}^2$ (d) $32.2\,\text{m}^3$

Qu. 4 Calculate the volume of a prism of rectangular cross-section 20 mm by 15 mm and length 100 mm.

Ans (a) $30\,\text{cm}^3$ (b) $300\,\text{mm}^3$
(c) $7600\,\text{mm}^2$ (d) $6000\,\text{mm}^3$

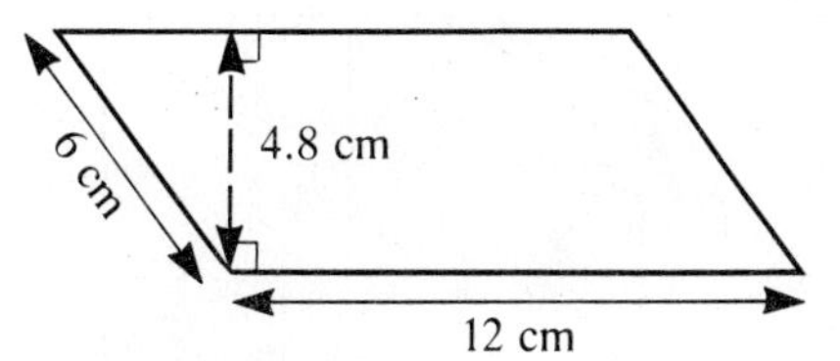

Figure 12.24 Diagram for Qu 1.

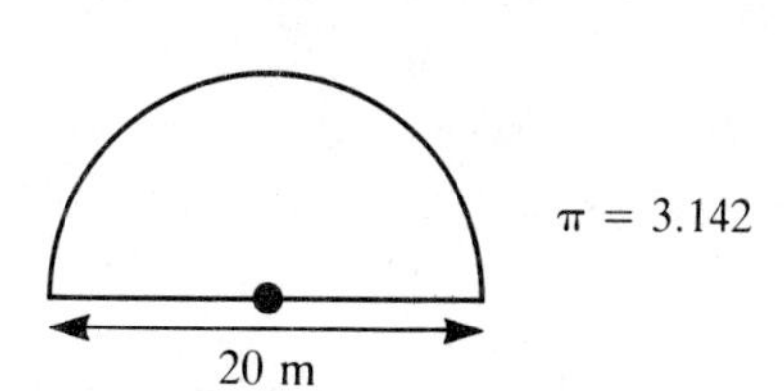

Figure 12.25 Diagram for Qu 2.

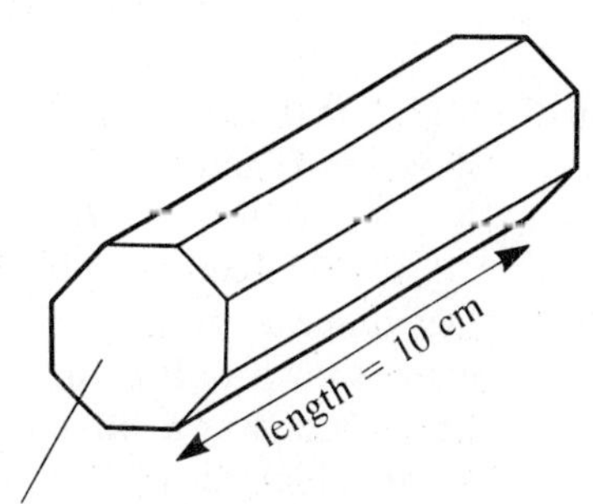

Figure 12.26 Diagram for Qu 6.

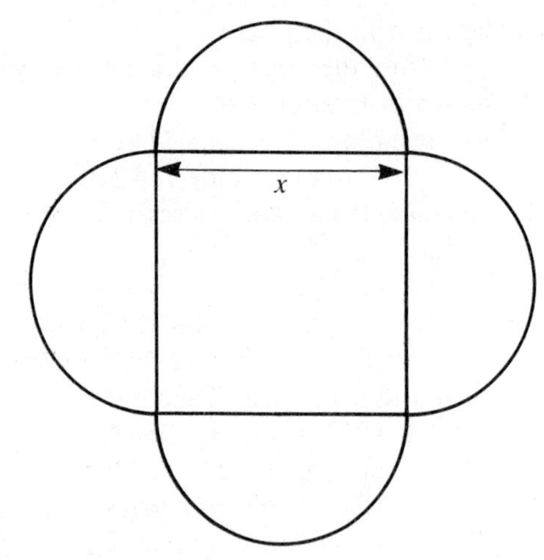

Figure 12.27 Diagram for Qu 7.

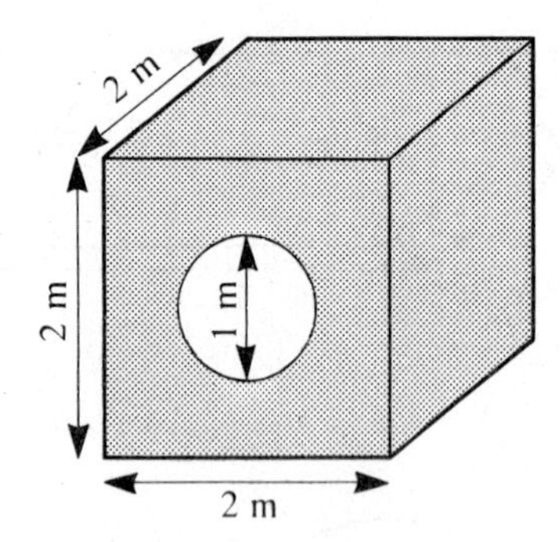

Figure 12.28 Diagram for Qu 8.

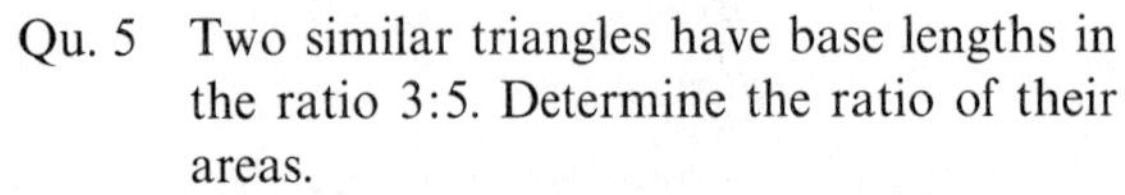

Qu. 5 Two similar triangles have base lengths in the ratio 3:5. Determine the ratio of their areas.

Ans (a) cannot be calculated due to insufficient information (b) 3:5 (c) 9:15 (d) 9:25

Qu. 6 Calculate the total surface area of the prism shown in Figure 12.26.

Ans (a) $86\,\text{cm}^3$ (b) $88.6\,\text{cm}^2$ (c) $97.2\,\text{cm}^2$ (d) $688\,\text{cm}^2$

Qu. 7 Determine an expression for the area of the figure shown in Figure 12.27, where semicircles are drawn on the four sides of a square of length x.

Ans (a) $x^2 + \pi x^2$ (b) $x^2 + 2\pi x^2$ (c) $x^2 + \frac{1}{2}\pi x^2$ (d) $2\pi x^3$

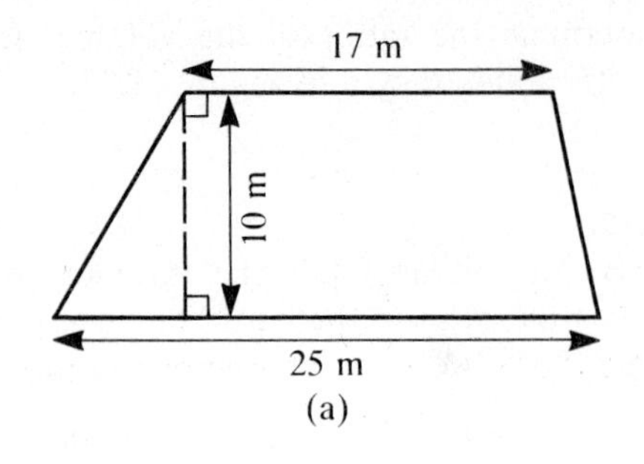

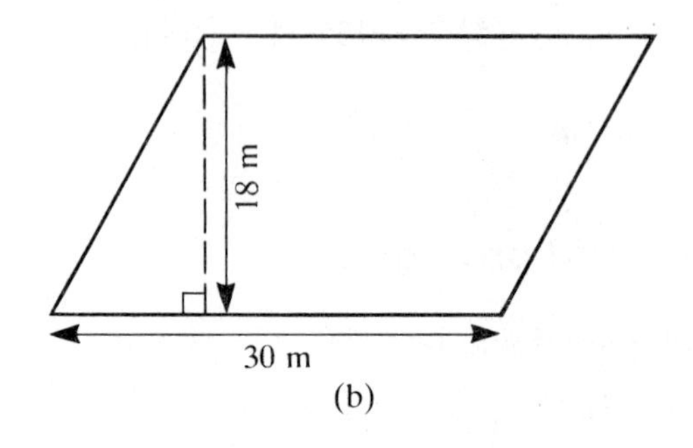

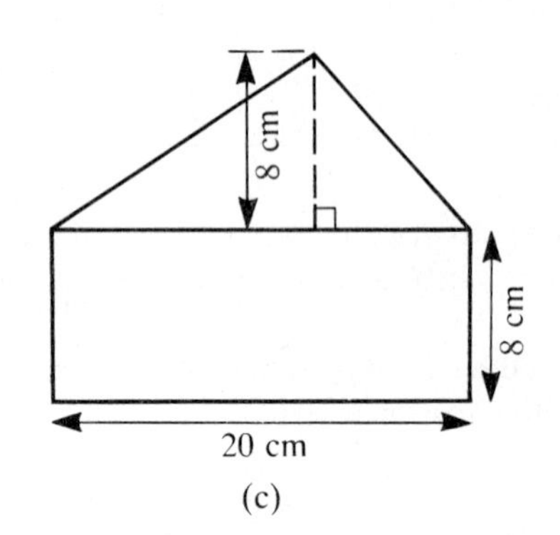

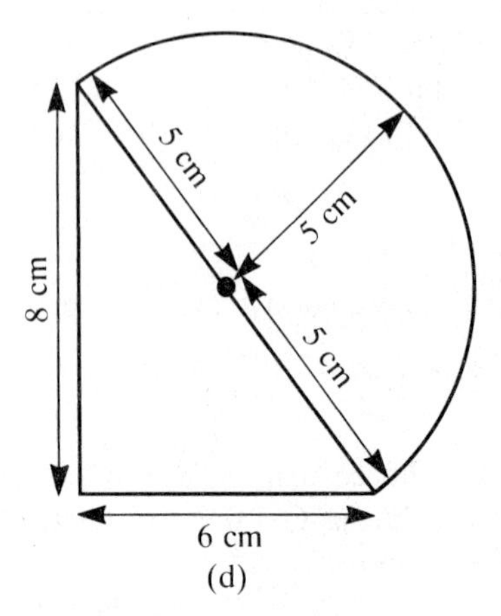

Figure 12.29 Figures for problem 1.

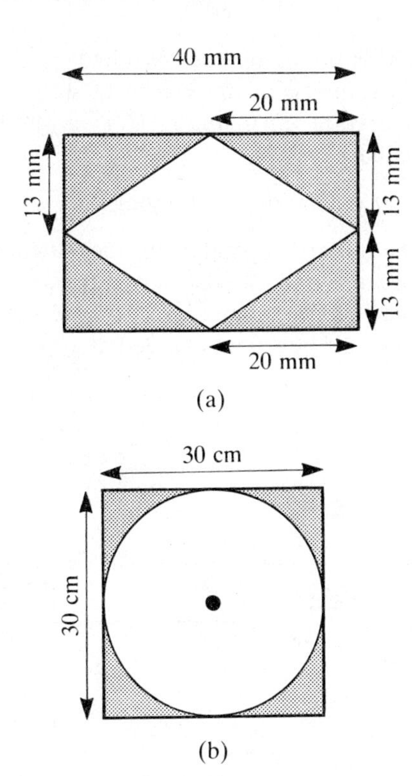

Figure 12.30 Areas for problem 2.

Qu. 8 A hole of circular cross-section and diameter 1 metre is cut completely through a 2 metre block of material, see Figure 12.28. Calculate the volume of the remaining material.

Ans (a) $6.43\,m^3$ (b) $4.86\,m^3$
(c) $0.79\,m^3$ (d) $1.57\,m^3$

Problems 12

1 Calculate the areas of the plane figures shown in Figure 12.29.

2 Calculate the areas shown shaded in Figure 12.30

3 Calculate the volumes and total surfaces areas for the solids shown in Figure 12.31.

4 Determine the volume and mass of
(a) a uniform steel rod of circular cross-section of radius 5 mm and length 300 mm; the density of steel is 7700 kilograms per cubic metre;
(b) air contained in a 'rectangular' room of dimensions 12 m by 8 m by 3 m; the density for air is $1.29\,kg/m^3$ (at 0°C and at standard atmospheric pressure, the conditions which are assumed present in the room).

5 Calculate the volume and total surface area of the V-block shown in Figure 12.32.

6 Figure 12.33 shows a diagram of the cross-section of a swimming pool taken along its length. The width of the pool is 15 m. Calculate the maximum capacity of the pool.

The pool is filled to within 10 cm of its free surface. Calculate the volume and mass of water now in the pool, given the density of water is $1000\,kg/m^3$.

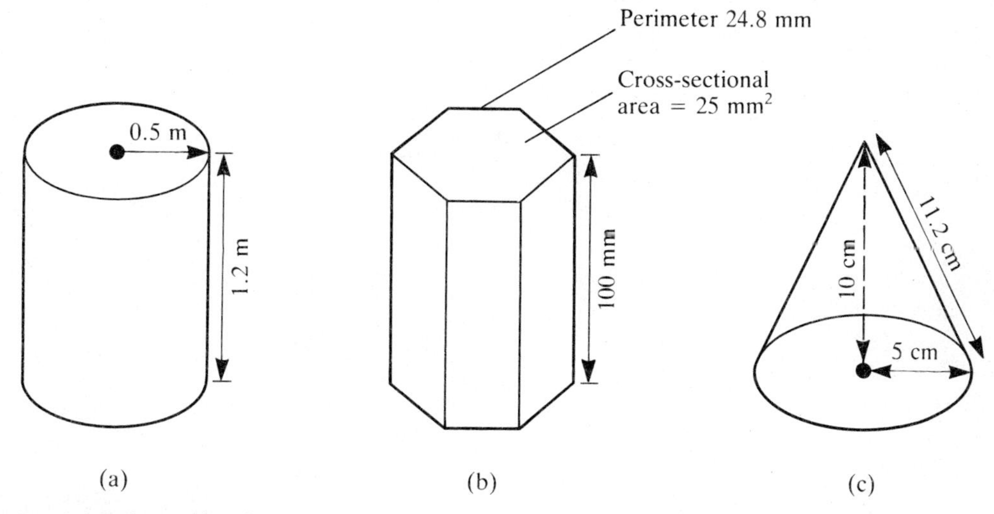

Figure 12.31 Solids for problem 3.

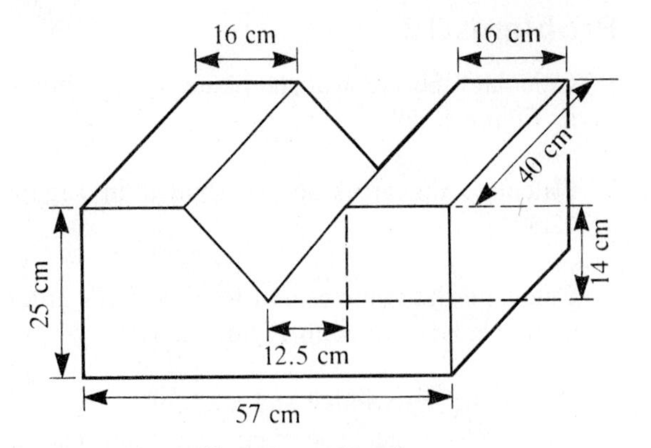

Figure 12.32 V-block for problem 5.

The amount of heat required to raise the temperature of a mass m kg of water from a temperature T_1°C to T_2°C is given by the formula:

$$H = 4200\,m(T_2 - T_1) \text{ joules}$$

If the initial temperature of the water is 10°C estimate H and the cost to heat the water in the pool to 24°C assuming 3.6×10^6 joules of energy (1 kilowatt hour) are charged at 10p.

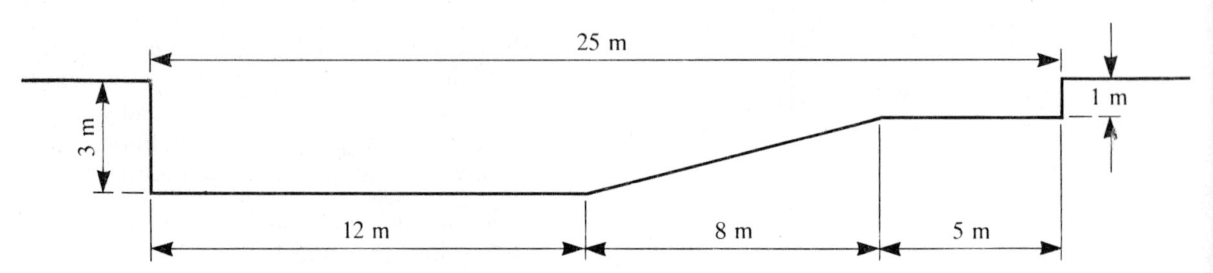

Figure 12.33 Cross-section of pool for problem 6.

13 Types and properties of triangles

General learning objectives: to recognize the types and properties of triangles.

13.1 Types of triangles: acute-angled, right-angled, obtuse-angled, equilateral and isosceles

A triangle is a plane figure bounded by three intersecting straight lines, as shown in Figure 13.1. The symbol △ is used as an abbreviation for triangle, and the symbols ∠ e.g. ∠A and ^, e.g. $B\hat{A}C$ are used to denote angles.

All triangles have the property that the sum of their internal angles is 180°, i.e. with reference to △ABC in Figure 13.1,

$$\angle A + \angle B + \angle C = 180^\circ$$

The names given to the various types of triangles defining their angle or side length characteristic are explained below.

An **acute-angled** triangle is one in which all internal angles are *acute*, that is less than 90°. Some examples of acute-angled triangles are shown in Figure 13.2.

A special case of an acute-angled triangle occurs when all three angles are equal, i.e. each internal angle equals $180 \div 3 = 60^\circ$. Such a triangle is known as an **equilateral triangle**. All three sides as well as the three angles are equal in an equilateral triangle.

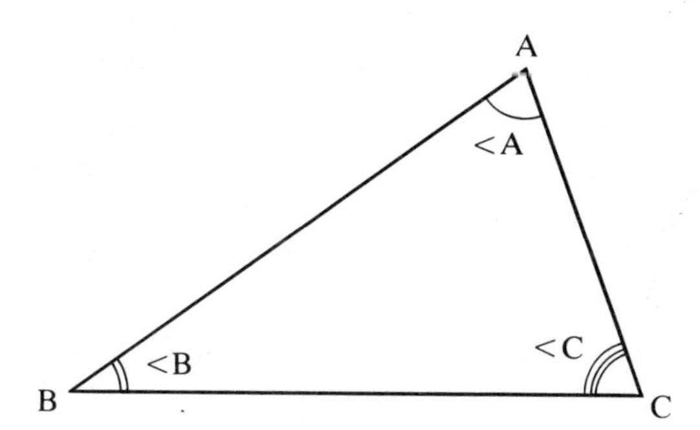

Figure 13.1 Triangle ABC, △ABC.

Some examples of equilateral triangles are shown in Figure 13.3.

A **right-angled triangle** as its name suggests has an angle equal to 90°. The longest side of a right-angled triangle, the side opposite the right-angle, is known as the **hypotenuse**. All right-angled triangles possess the property: 'the square on the hypotenuse equals the sum of the squares on the other two sides' – Pythagoras' theorem, which we consider in section

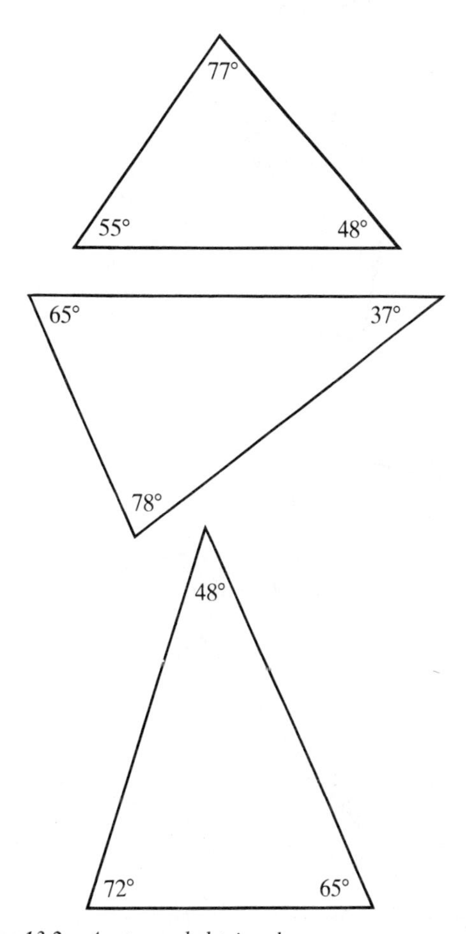

Figure 13.2 Acute-angled triangles.

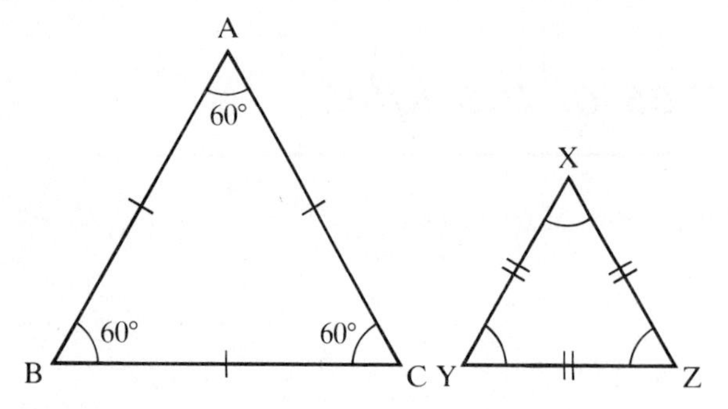

Figure 13.3 Equilateral triangles (all sides equal, all angles equal) Note: equal sides are normally denoted on diagrams by oblique stroke(s) on the actual lines, e.g. + or ⧺; equal angles can also be denoted by ∢ or ∢∢.

13.3. Some examples of right-angled triangles are shown in Figure 13.4.

An obtuse-angled triangle is one which contains an angle greater than 90° and obviously less than 180°, i.e. an obtuse angle. Some examples of obtuse-angled triangles are shown in Figure 13.5.

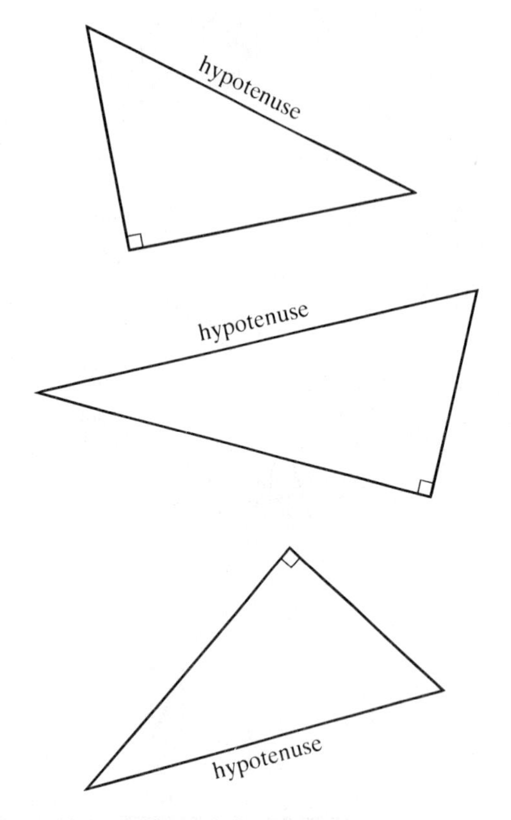

Figure 13.4 Right-angled triangles Note: the right-angle, 90° angle, is denoted by ⦜.

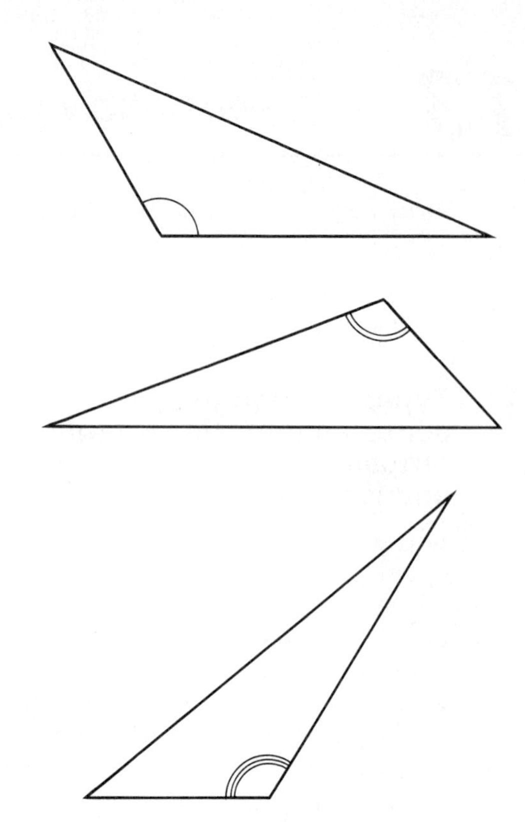

Figure 13.5 Obtuse-angled triangles.

Triangles which have two equal sides are known as **isosceles** triangles. In an isosceles triangle the internal angles contained by the equal length sides with the third side are also equal. Some examples of isosceles triangles are shown in Figure 13.6.

Triangles in which all three sides have different lengths are known as **scalene** triangles.

13.2 Angle properties of a triangle and complementary angles

The sum of the internal angles in a triangle equals 180°. Also if any side of a triangle is extended, the angle formed with the adjacent side of a triangle is known as the exterior angle and we have the general property for triangles:

exterior angle = sum of two internal opposite angles

i.e. in Figure 13.7,

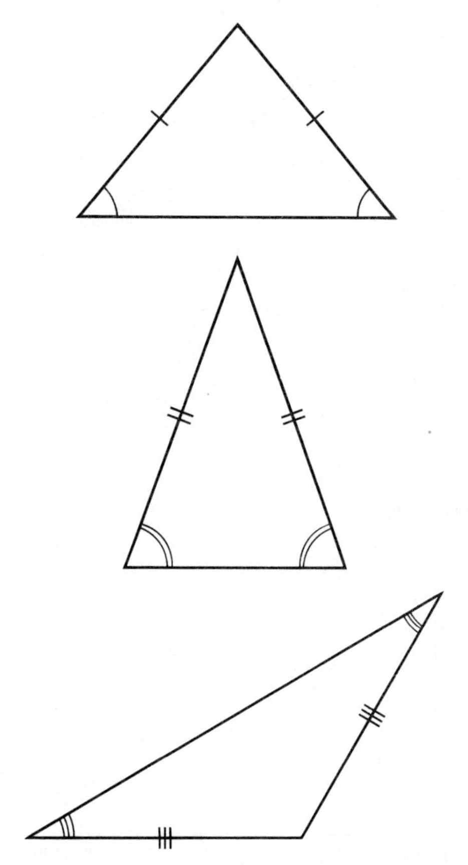

Figure 13.6 Isosceles triangles (two sides equal).

$$\text{exterior angle, } A\hat{C}X = \text{sum of internal opposite angles} = \angle A + \angle B$$

These two fundamental results for a triangle can easily be proved by reference to Figure 13.7(b) where the line CD, parallel to side BA is also drawn. Using

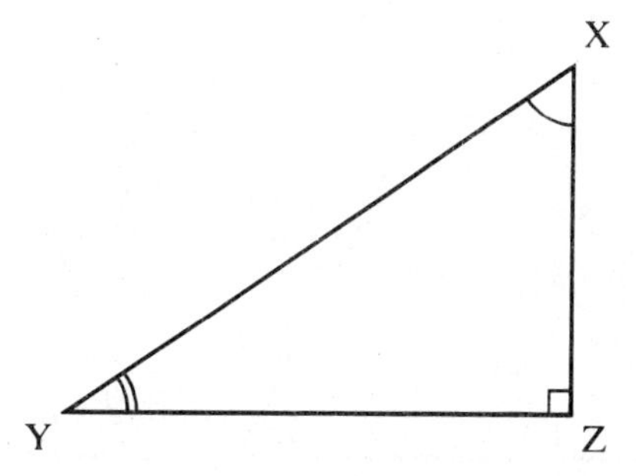

Figure 13.8 Angles $\angle X$ and $\angle Y$ in a right-angled triangle are complementary, they sum up to 90°.

the property of parallel lines cut by transversal (diagonal line), we obtain

$$\angle A = A\hat{C}D \quad (\text{alternate } \angle s,\ AB \parallel CD)$$
$$\angle B = D\hat{C}X \quad (\text{corresponding } \angle s,\ AB \parallel CD)$$

so on adding,

$$\angle A + \angle B = A\hat{C}X$$

i.e. sum of two internal opposite angles = exterior angle

Also $\angle C + A\hat{C}X = 180°$ (angles in straight line sum to 180°)
and as $A\hat{C}X = \angle A + \angle B$
we have $\angle C + \angle A + \angle B = 180°$

i.e. sum of internal angles in triangle equal 180°.

In a right-angled triangle as one internal angle is 90° the other two angles must sum to $180° - 90° = 90°$. When two angles sum to 90° they are said to be **complementary**, so in Figure 13.8,

$$\angle X + \angle Y = 90°$$

and the angles $\angle X + \angle Y$ are complementary

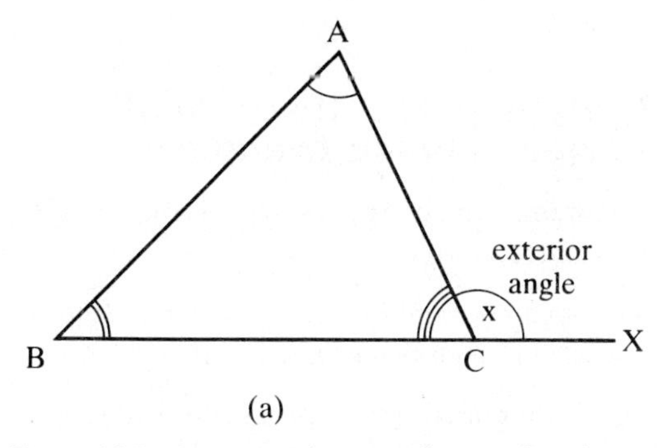

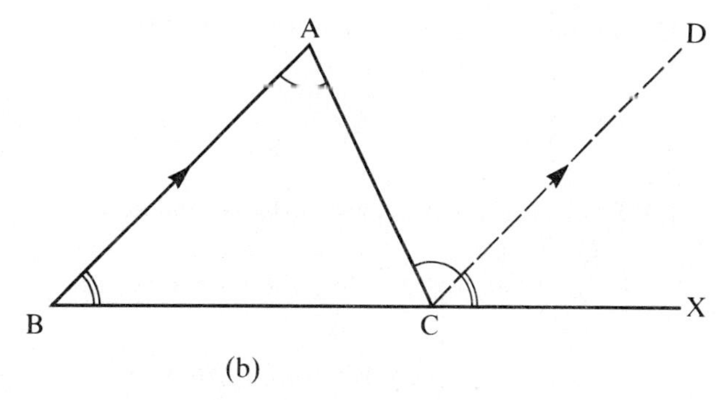

Figure 13.7 In a triangle: sum of internal angles sum to 180° exterior angle = sum two internal opposite angles (a) (b).

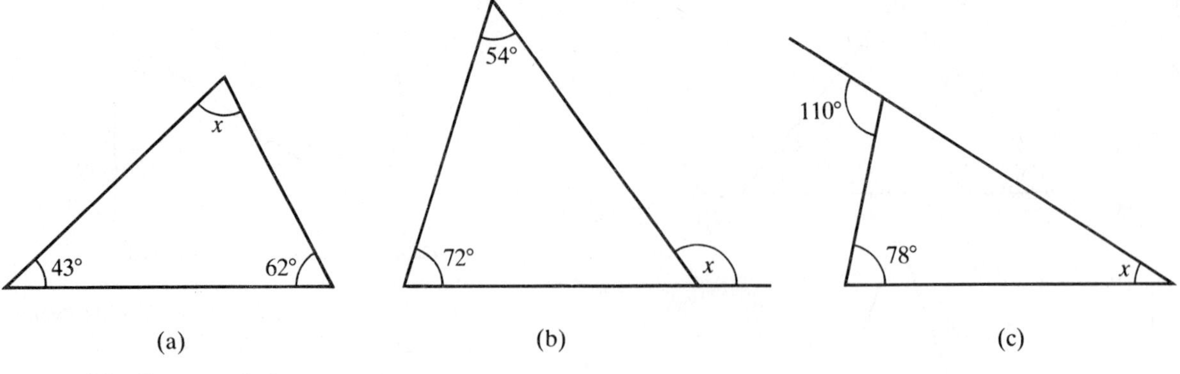

Figure 13.9 For example 1.

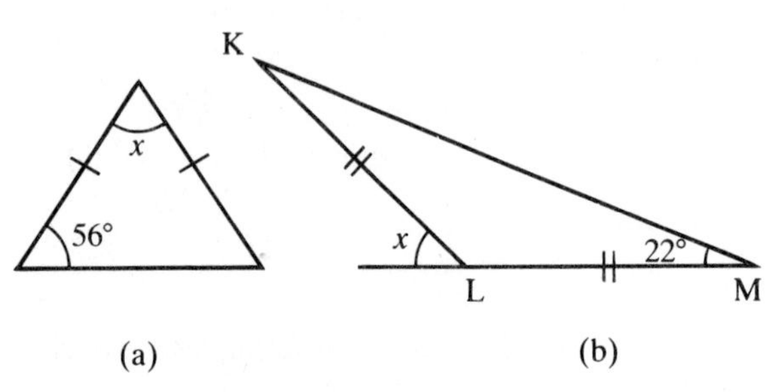

Figure 13.10 For example 2.

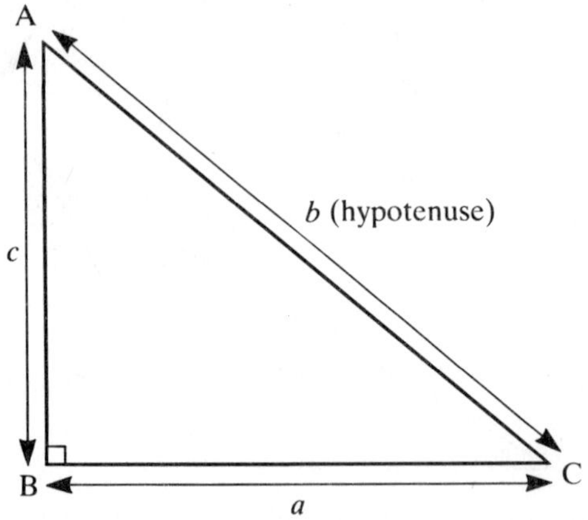

Examples

1 Determine the value of the angles x in Figure 13.9.

Solution

(a) $x = 180° - 43° - 62° = 75°$ *Ans*

(b) Since sum of internal opposite angles = exterior angle,

$x = 72° + 54° = 126°$ *Ans*

(c) Likewise, $78° + x = 110°$

so $x = 110° - 78° = 32°$ *Ans*

2 Determine the value of the angles x in Figure 13.10.

Solution

(a) The triangle is isosceles so the second base angle is also 56°. Hence, since the sum of the internal angles equals 180°, we have

$$56° + 56° + x = 180°$$
$$x = 180° - (2 \times 56) = 68°$$

Ans

Figure 13.11 Pythagoras' theorem for right-angled triangles: $b^2 = a^2 + c^2$.

(b) Since the triangle is isosceles,

$$\angle K = \angle M = 22°$$

and as the exterior angle = sum of two internal opposite angles,

$$x = 22° + 22° = 44° \quad Ans$$

13.3 Pythagoras' theorem of right-angled triangles

This theorem, which applies to all right-angled triangles states:

The square on the hypotenuse is equal to the sum of the squares on the other two sides.

Stating Pythagoras' theorem algebraically with reference to Figure 13.11, we have:

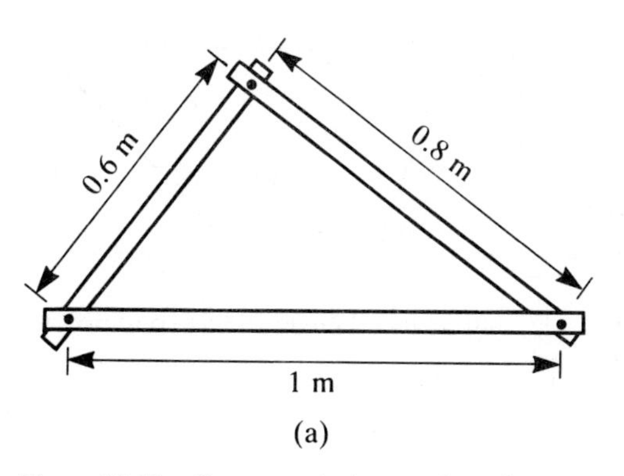

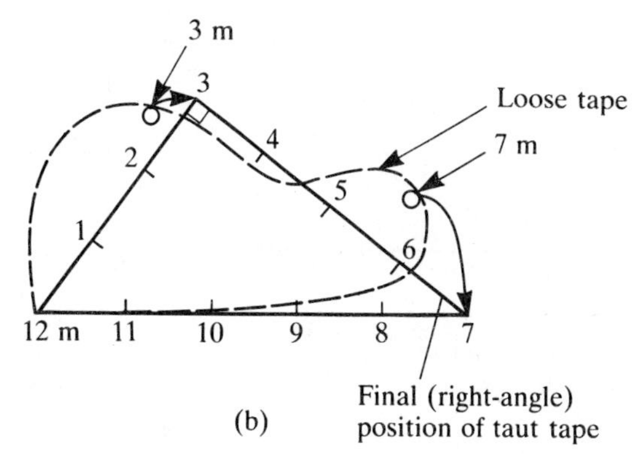

Figure 13.12 Two practical examples of constructing a right angle based on the 3:4:5 triangle.

$$b^2 = a^2 + c^2$$

square on hypotenuse (b^2); sum of squares on other two sides ($a^2 + c^2$)

The theorem is extremely useful for calculating lengths in right-angled triangles, given any two lengths. For example, to find the hypotenuse b when a and c are given, we have

$$b^2 = a^2 + c^2$$
$$b = \sqrt{(a^2 + c^2)}$$

Making a the subject of the formula,

$$a^2 = b^2 - c^2$$

so $\quad a = \sqrt{(b^2 - c^2)}$

Likewise, $\quad c = \sqrt{(b^2 - a^2)}$

There are some special cases of right-angled triangles whose sides are whole numbers. The best known is the 3, 4, 5 triangle where the lengths are in the ratio 3:4:5. For example,

if $\quad a = 3,\ c = 4,\ b = 5,$

then $\quad a^2 + c^2 = 3^2 + 4^2 = 9 + 16 = 25$

$\quad b^2 = 5^2 = 25$

so $b^2 = a^2 + c^2$ and the triangle is right-angled, with the right-angle opposite the hypotenuse, $b = 5$ length

Any multiple or sub-multiple of 3, 4, 5 forms a right-angled triangle, e.g.

the lengths 6, 8, 10 or 15, 20, 25 or 0.3, 0.4, 0.5 all form right-angled triangles.

A second well-known case is that triangles with lengths in the ratio 5:12:13 are right-angled triangles, as

$$5^2 + 12^2 = 25 + 144 = 169 = 13^2$$

The 3:4:5 triangle is often used in surveying and construction engineering as a simple means of obtaining an accurate right angle. For example, if wooden batterns are nailed together to form a

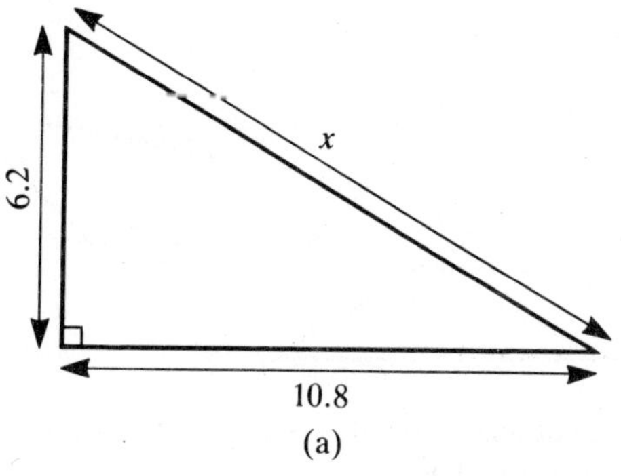

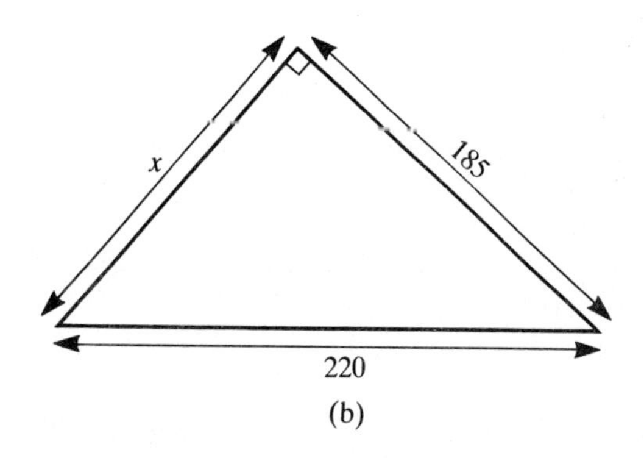

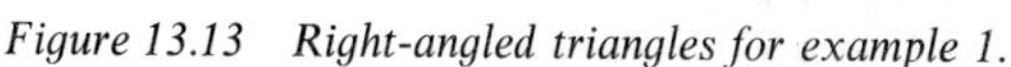

Figure 13.13 Right-angled triangles for example 1.

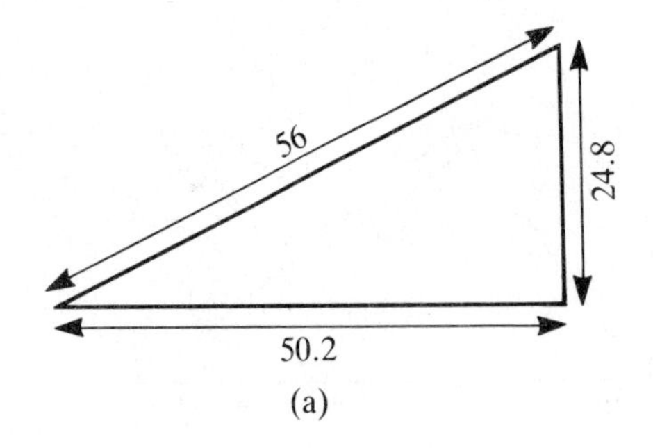

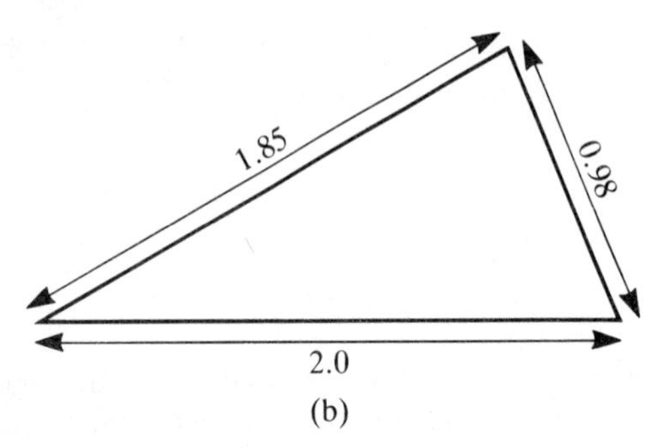

Figure 13.14 Triangles for example 2.

triangle with lengths of, say, 0.6 m, 0.8 m and 1 m between nailing points (see Figure 13.12(a)), the angle opposite the 1 m side is 90°. Alternatively a linen measuring tape may be employed. If the zero and 12 m mark, say, are placed together, and then the 12 m loop pulled into a triangle with the 3 m mark and the $(3 + 4) = 7$ m marks forming the other two vertices (corners) of the triangle, the angle at the 3 m mark is a right angle (see Figure 13.12(b)).

Examples

1 Determine by applying Pythagoras' theorem the side length marked x in the right-angled triangles of Figure 13.13.

Solution

(a) $x^2 = 6.2^2 + 10.8^2 = 38.44 + 116.64 = 155.08$

$x = \sqrt{155.08} = 12.45$ *Ans*

(b) $185^2 + x^2 = 220^2$

so $x^2 = 220^2 - 185^2 = 48\,400 - 34\,225 = 14\,175$

$x = \sqrt{14\,175} = 119.1$ *Ans*

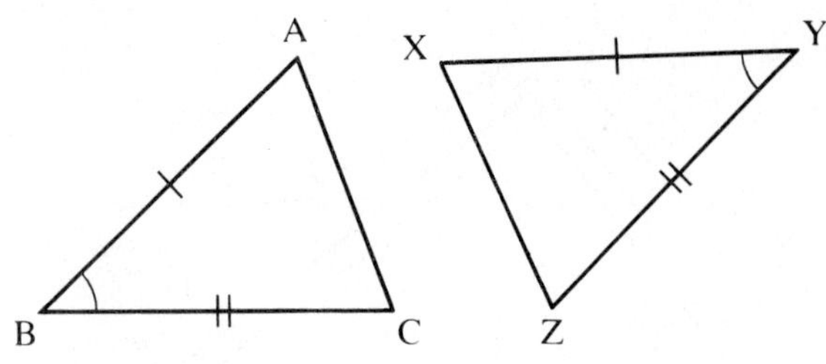

(a) Two sides and included angle (SAS) condition for congruency:
$\Delta ABC \equiv \Delta XYZ$ (SAS)

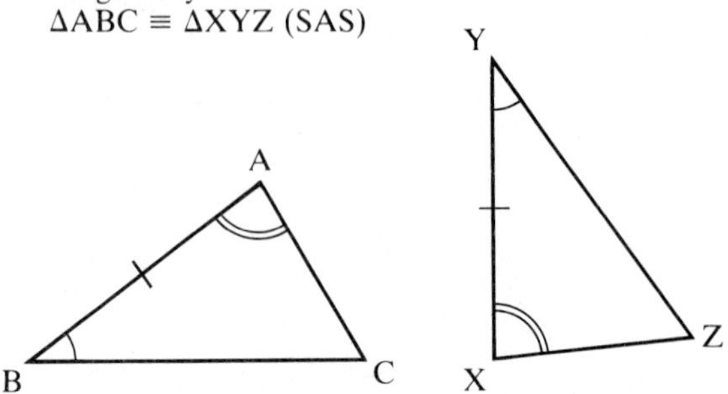

(b) (i) Angle, angle corresponding side (AAS) condition for congruency:
$\Delta ABC \equiv \Delta XYZ$ (AAS)

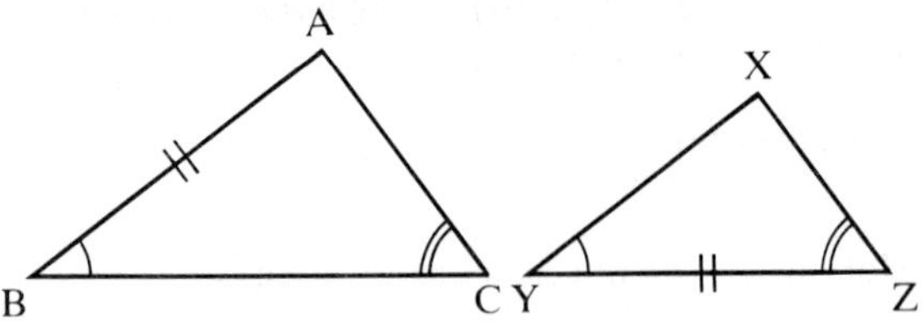

(ii) Case where two angles and a side are equal, but clearly triangles are not congruent

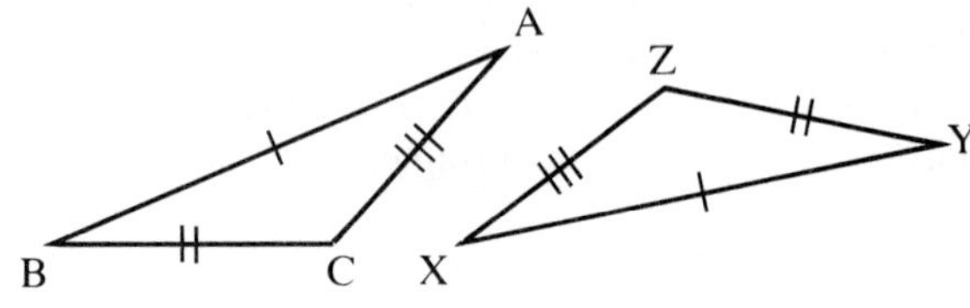

(c) All 3 sides equal (SSS) condition for congruency:
$\Delta ABC \equiv \Delta XYZ$ (SSS)

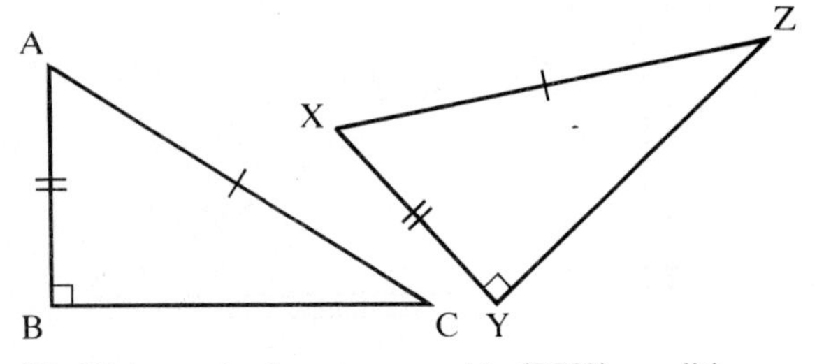

(d) Right-angle, hypotenuse, side (RHS) condition for congruency
$\Delta ABC \equiv \Delta XYZ$ (RHS)

Figure 13.15 The four conditions for congruency: test conditions used to prove one triangle is identical to another.

2 Determine using Pythagoras' theorem whether or not the triangles in Figure 13.14 are right-angled triangles.

Solution

If the triangles are right-angled then the square the longest side should equal the sum of the squares of the other two sides. Hence

(a) $56^2 = 3136$
$24.8^2 + 50.2^2 = 615.0 + 2520.0 = 3135$
so within the limits of accuracy, the square of the 56 hypotenuse equals the sum of the squares of 24.8 and 50.2 sides, hence the triangle is right-angled.

(b) $2^2 = 4$
$1.85^2 = 3.42$, $0.98^2 = 0.96$
so $1.85^2 + 0.98^2 = 3.42 + 0.96 = 4.38 \neq 4$
i.e. Pythagoras' theorem is not satisfied, so the triangle does not contain a right-angle.

13.4 Congruent triangles: conditions for triangles to be identical

If two triangles are identical in all respects (identical length sides and angles) they are said to **congruent** triangles. Congruency is expressed mathematically using the ≡ identical sign, e.g.

$$\triangle ABC \equiv \triangle XYZ$$

expresses the fact the triangles ABC and XYZ are congruent.

There are four conditions that may be used to prove congruency and these are summarized in Figure 13.15.

Example

State, giving reasons, whether or not the pairs of triangles shown in Figure 13.16 are congruent.

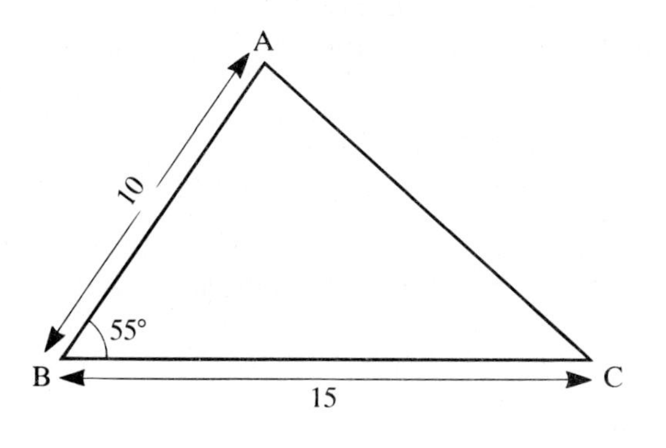

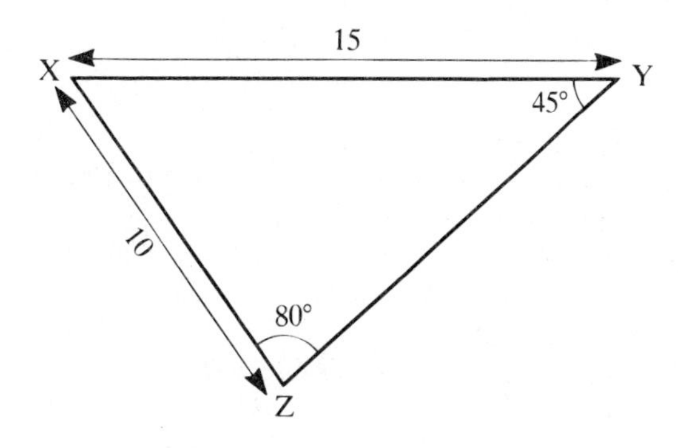

(a)

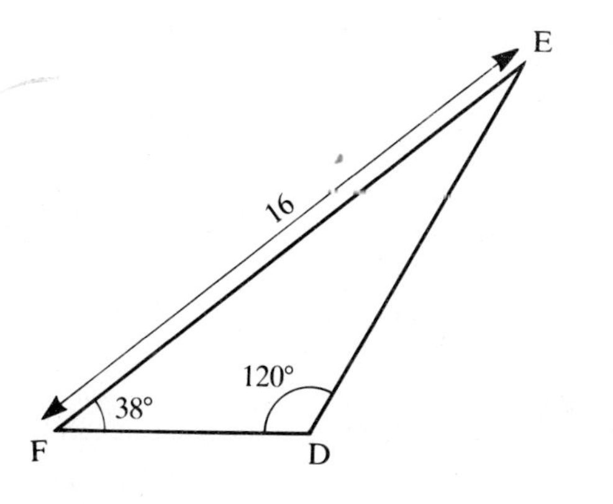

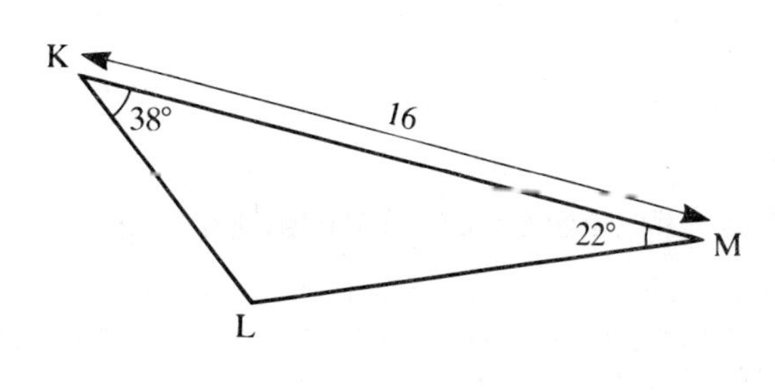

(b)

Figure 13.16 Triangles for congruency test examples.

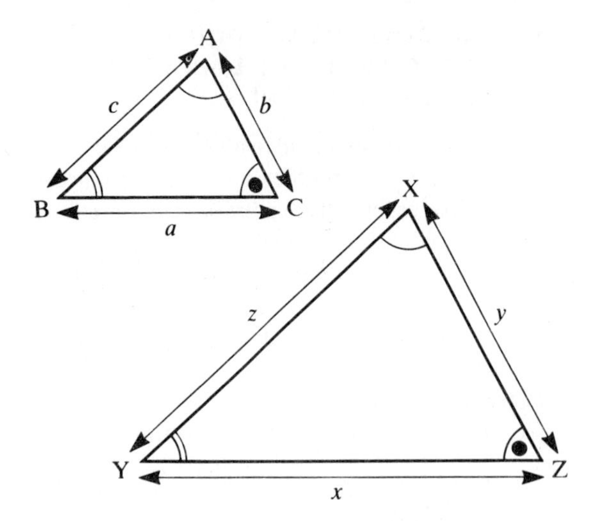

Figure 13.17 Similar triangles (angles of one respectively equal to angles of another). Similar triangles possess the property that the ratio of corresponding sides are equal: $\frac{a}{x} = \frac{b}{y} = \frac{c}{z}$.

Solution

(a) In $\triangle$ABC and $\triangle$XYZ of Figure 13.16(a),

AB = XZ = 10
BC = XY = 15
$\angle$X = 180° − 80° − 45° = 55°

so $\angle$B = $\angle$X

hence $\triangle$ABC ≡ $\triangle$XYZ (SAS)

(b) In $\triangle$EFD and $\triangle$KLM of Figure 13.16 (b)

EF = KM = 16 (sides corresponding to opposite $\angle$D and $\angle$L)
$\angle$F = $\angle$K = 38°
$\angle$L = 180° − 38° − 22° = 60°

so $\angle$D = $\angle$L

hence $\triangle$EFD ≡ $\triangle$KLM (AAS)

13.5 Properties of similar triangles

Triangles in which the individual angles are respectively equal to the angles in another are known as **similar triangles**. They have similar shape but not in general equal sides. Only when the triangles are congruent will all corresponding sides and angles be equal. Triangles ABC and XYZ in Figure 13.17 are similar triangles, since

$$\angle A = \angle X, \quad \angle B = \angle Y, \quad \angle C = \angle Z$$

In fact it is only necessary to prove or be given that two angles of one triangle are respectively equal to two angles of another to state that the two triangles are similar, since the third will be equal automatically as the three angles sum to 180°.

Similar triangles have the very important property that the ratio of corresponding sides are equal, i.e. with reference to Figure 13.17

$$\frac{a}{x} = \frac{b}{y} = \frac{c}{z}$$

The qualification of 'corresponding' is used to denote that the ratios must be formed by taking sides opposite the equal, i.e. corresponding, angles. Side a is opposite $\angle$A, side x is opposite $\angle$X which is the corresponding angle to $\angle$A. Likewise b and y are corresponding sides, opposite the equal angles $\angle$B and $\angle$Y; and c and z are corresponding sides.

The ratio equality property of similar triangles is very useful and is used frequently to determine unknown lengths in problems where similar triangles occur.

Examples

1 In Figure 13.18 RS is parallel to BC, AB = 6, AC = 7.5 and AR = 3.8. Determine the lengths AS and SC

Solution

Since RS ∥ BC, $\triangle$ARS and $\triangle$ABC are similar. Hence applying the equal ratio property

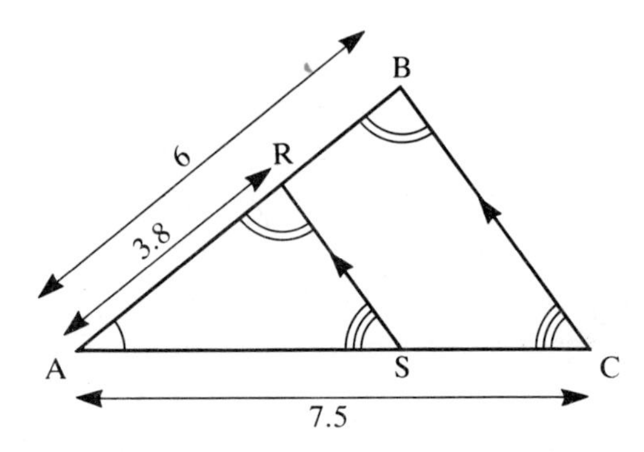

Figure 13.18 For example 1.

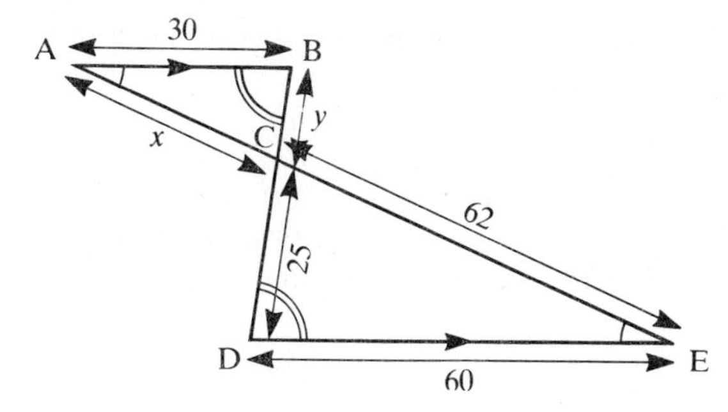

Figure 13.19 For example 2.

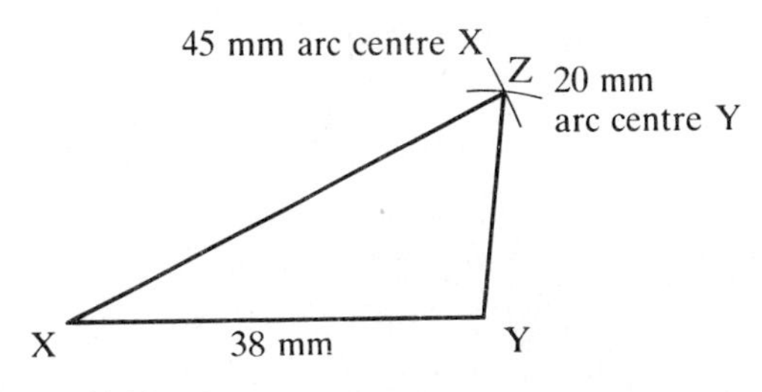

Figure 13.20 Construction of triangle in which $XY = 38$ mm, $YZ = 20$ mm, $XY = 45$ mm (see example 1).

$$\frac{AR}{AB} = \frac{3.8}{6} = \frac{AS}{AC} = \frac{AS}{7.5}$$

i.e. $\frac{3.8}{6} = \frac{AS}{7.5}$

so $AS = \frac{3.8 \times 7.5}{6} = 4.75$ *Ans*

and $SC = AC - AS = 7.5 - 4.75 = 2.75$ *Ans*

2 Calculate the lengths x and y in Figure 13.19.

Solution

Since $\triangle$ABC and $\triangle$CDE are similar, we have

$$\frac{x}{62} = \frac{30}{60} = \frac{y}{25}$$

so $\frac{x}{62} = \frac{30}{60} = \frac{1}{2}$, hence $x = \frac{1}{2} \times 62 = 31$ *Ans*

$\frac{y}{25} = \frac{30}{60} = \frac{1}{2}$, hence $y = \frac{1}{2} \times 25 = 12.5$ *Ans*

13.6 Construction of triangles from given data

Triangles can be constructed from given data, i.e. information of lengths and/or angles using a rule, protractor and compass. Although limited information may be given it must be sufficient to define a unique triangle. Triangles may be uniquely defined by one of the four congruency conditions:

1 side, included angle, side (SAS)
2 Angle, angle, corresponding side (AAS)
3 Side, side, side (SSS)
4 Right-angle, hypotenuse, side (RHS)

Any other specification would lead to more than one possible triangle. For example, if three angles only were given there would be an infinite number of similar triangles; if two sides are given and an angle, which is not 'included' by the sides, there are two possible solutions.

The construction of triangles is illustrated in the following examples.

Examples

1 Construct triangle XYZ in which $XY = 38$ mm, $YZ = 20$ mm and $XZ = 45$ mm.
Measure the angles $\angle X$, $\angle Y$ and $\angle Z$

Construction

(i) Draw line $XY = 38$ mm.
(ii) Using a compass with Y as centre to draw an arc of radius $YZ = 20$ mm.
(iii) With X as centre draw a second arc of radius $XZ = 45$ mm.
The point of intersection of the two arcs gives point Z
(iv) Finally join XZ and YZ by straight lines

The construction is shown in Figure 13.20.
Measuring the angles we obtain $\angle X = 25°$, $\angle Y = 98°$, $\angle Z = 57°$

2 Construct the triangle KLM in which $KL = 40$ mm, $KM = 35$ mm and $\angle K = 50°$. Measure length LM and angles $\angle L$ and $\angle M$.

Construction

(i) Draw line $KL = 40$ mm.
(ii) Use a protractor to construct $\angle K = 50°$ at point K.

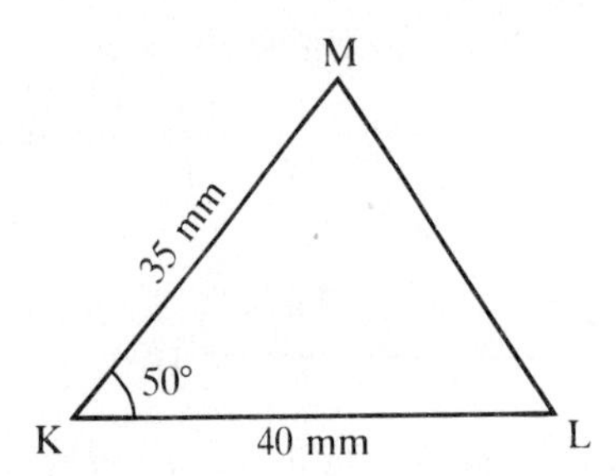

Figure 13.21 Construction triangle given $KL = 40$ mm, $KM = 35$ mm, $K = 50°$ (see example 2).

(iii) Mark off point M on the 50° line so KM = 35 mm.
(iv) Joint points LM.

The construction and complete △KLM is shown in Figure 13.21. From measurements, LM = 32 mm, ∠L = 57°, ∠M = 73°.

3 Construct a right-angled triangle RST with hypotenuse RS = 50 mm and side ST = 25 mm. Measure side RT and check the result using Pythagoras' Theorem.

*Construction**

(i) First draw the hypotenuse RS = 50 mm and locate centre of RS, point M in Figure 13.22.
(ii) Using a compass draw a semicircle centre M with RS as diameter, i.e. of radius $\frac{1}{2} \times 50 = 25$ mm.
(iii) With S as centre draw in arc of radius ST = 25 mm.
The point of intersection of this arc with the semicircle is point T.
(iv) Join RT and ST by straight lines.

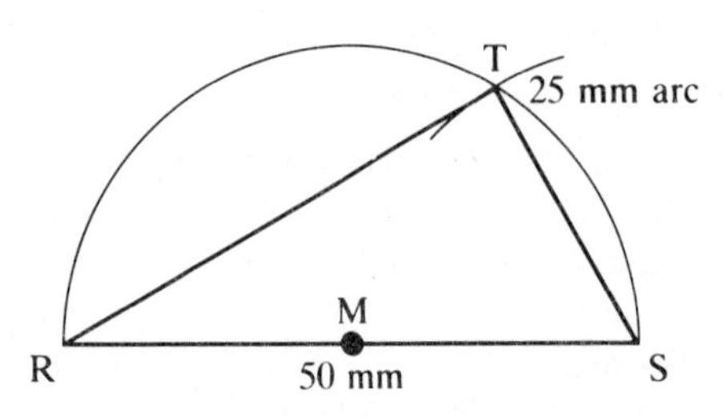

Figure 13.22 Construction of right-angled triangle RST in which hypotenuse $RS = 50$ mm, and side $ST = 25$ mm (see example 3).

From measurement on Figure 13.22, RT = 43 mm.

Check using Pythagoras' theorem

$$RT^2 = RS^2 - ST^2 = 50^2 - 25^2$$
$$= 2500 - 625 = 1875$$

so $RT = \sqrt{1875} = 43.3$ mm

Test and problems 13

Multiple choice test: MT 13

Answer block:

Question No.	0	1	2	3	4	5	6	7	8
Answer	d								

Enter your answer, that is a, b, c or d in the column under the question number in the answer block above. Note that question Qu. 0 has already been worked out and the answer inserted.

Qu. 0 Determine the length x in the right-angled triangle of Figure 13.23.

Ans (a) 4 (b) 10 (c) 13 (d) 12

Solution

Using Pythagoras' theorem,

$$20^2 = 16^2 + x^2$$

so $x^2 = 20^2 - 16^2 = 400 - 256 = 144$

$x = \sqrt{144} = 12$ *Ans*

Hence the answer is (d) and 'd' is inserted in the answer block under Qu. 0 as shown.

**Note*: In Chapter 14 where we consider the properties of circles we show that any angle subtended by the diameter on a semicircle is a right angle. Thus by constructing a semicircle on RS we know that point T must lie on the semicircle. The actual point is located by step (ii) above.

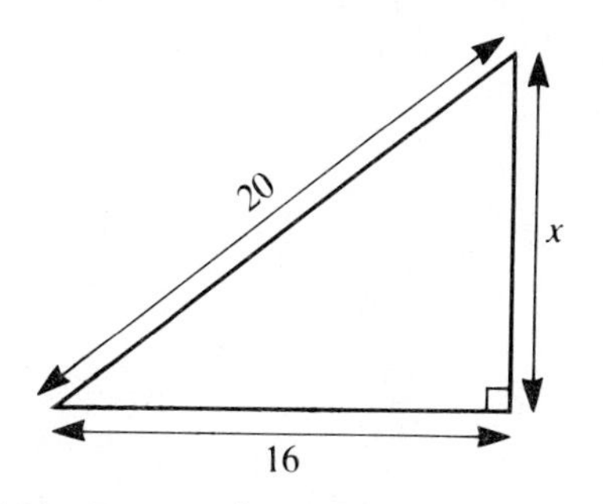

Figure 13.23 Diagram for Qu 0.

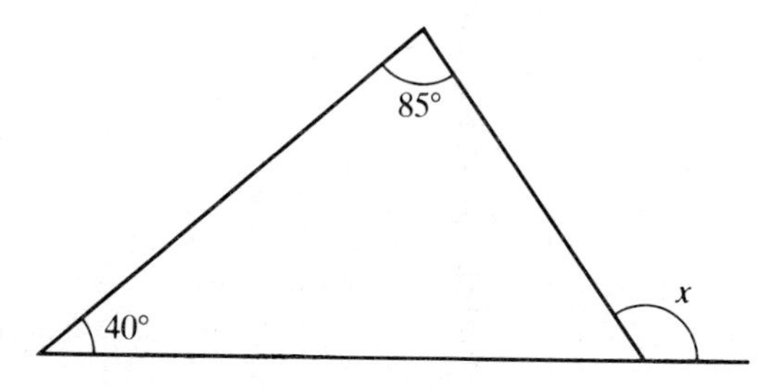

Figure 13.24 Diagram for Qu 1.

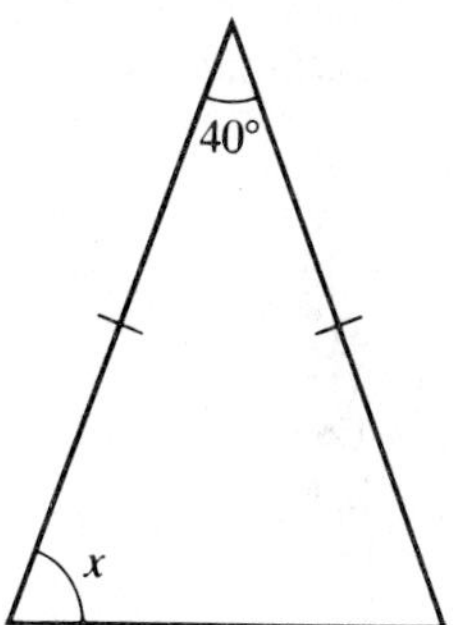
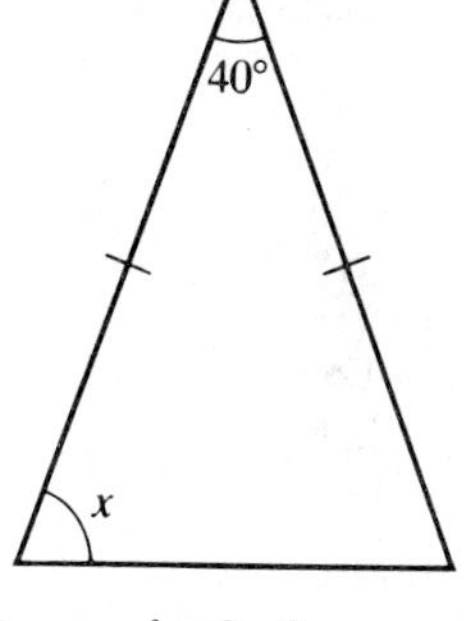

Figure 13.25 Diagram for Qu 2.

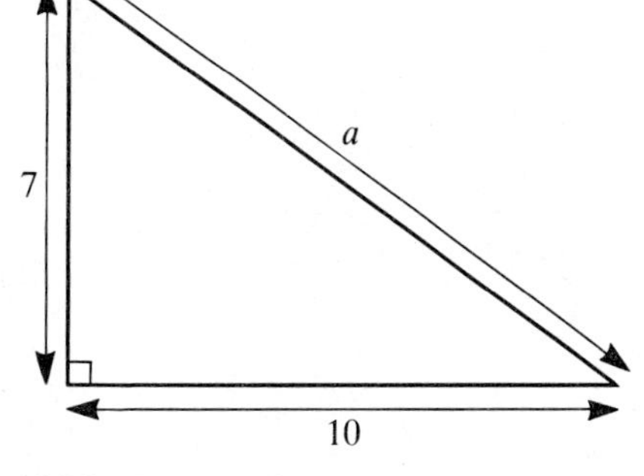

Figure 13.26 Diagram for Qu 3.

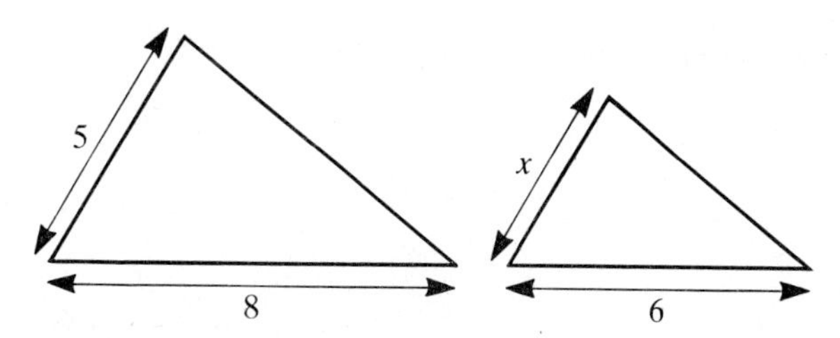

Figure 13.27 Diagram for Qu 4.

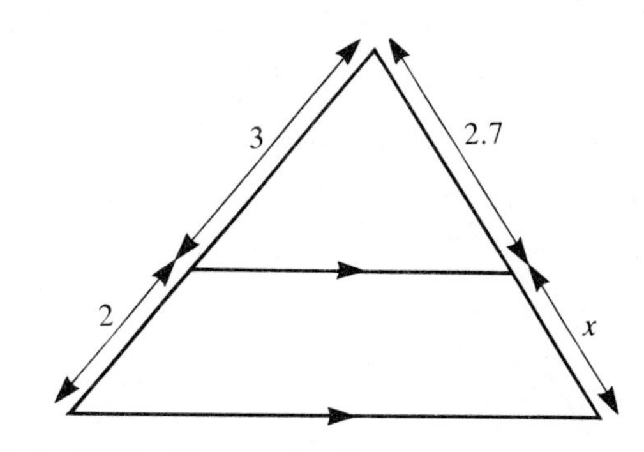

Figure 13.28 Diagram for Qu 7.

Now carry on with the test.

Qu. 1 Determine the angle x in the diagram of Figure 13.24.
Ans (a) 85° (b) 40° (c) 105° (d) 125°

Qu. 2 Determine the angle x in Figure 13.25.
Ans (a) 70° (b) 40°
(c) 140° (d) 60°

Qu. 3 Determine the length a in the triangle of Figure 13.26.
Ans (a) 17 (b) $\sqrt{149}$
(c) 11 (d) 13.52

Qu. 4 The triangles in Figure 13.27 are similar. Determine the length x.
Ans (a) 3.75 (b) 4 (c) 3.5 (d) 2.9

Qu. 5 Which of the following conditions cannot be used to prove congruency?
(a) angle, angle, corresponding side (AAS)
(b) side, angle, side (SAS)
(c) angle, angle, angle (AAA)
(d) right-angle, hypotenuse, side (RHS)

Qu. 6 The side lengths of four triangles are given below in the answer. Which one is a right-angled triangle?
Ans (a) 2, 3, 4 (b) 6, 8, 10
(c) 7, 8, 11 (d) 5, 13, 15

Qu. 7 Determine the value of x in the diagram of Figure 13.28.

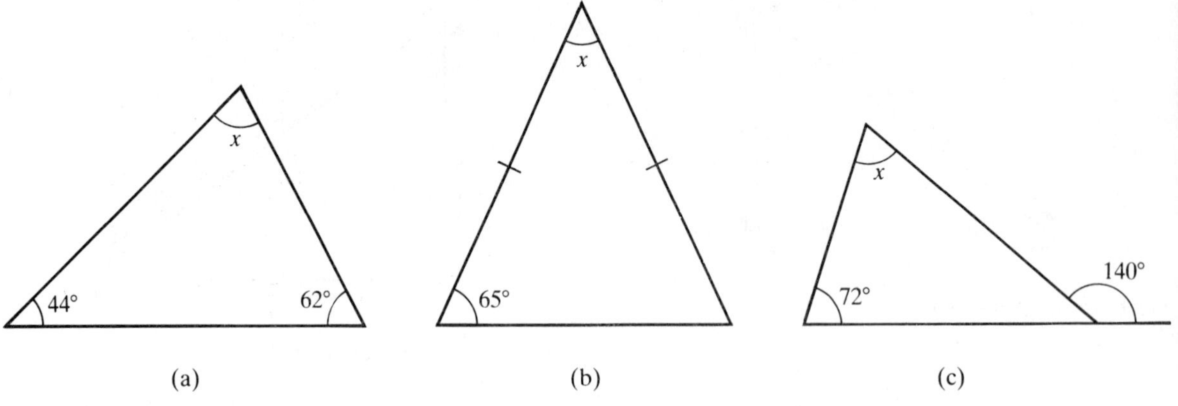

Figure 13.29 Diagrams for problem 1.

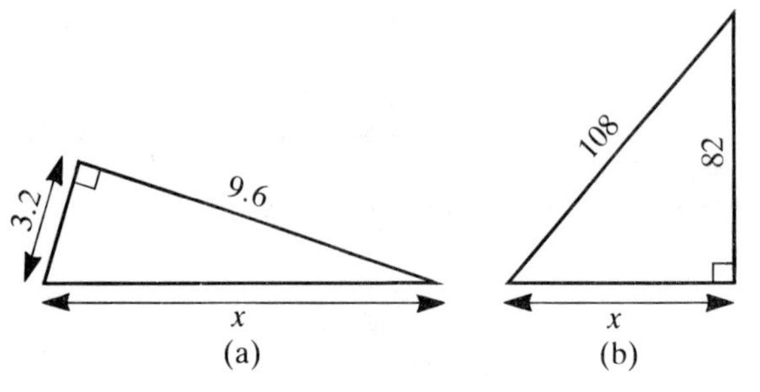

Figure 13.30 Diagrams for problem 2.

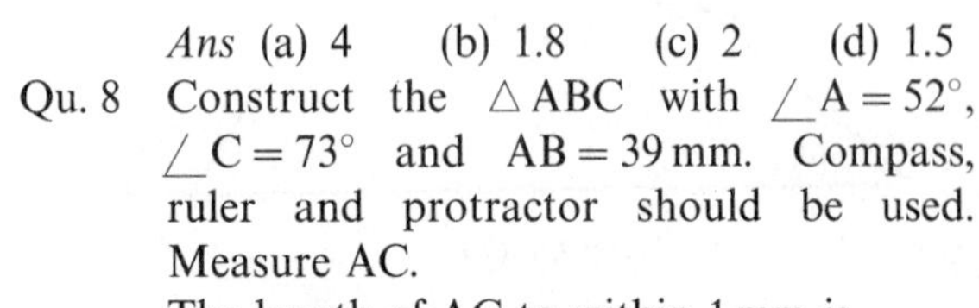
Ans (a) 4 (b) 1.8 (c) 2 (d) 1.5

Qu. 8 Construct the $\triangle$ABC with $\angle A = 52°$, $\angle C = 73°$ and AB = 39 mm. Compass, ruler and protractor should be used. Measure AC.
The length of AC to within 1 mm is
Ans (a) 25 mm (b) 47 mm
(c) 15 mm (d) 33 mm

Figure 13.31 Diagrams for problem 3.

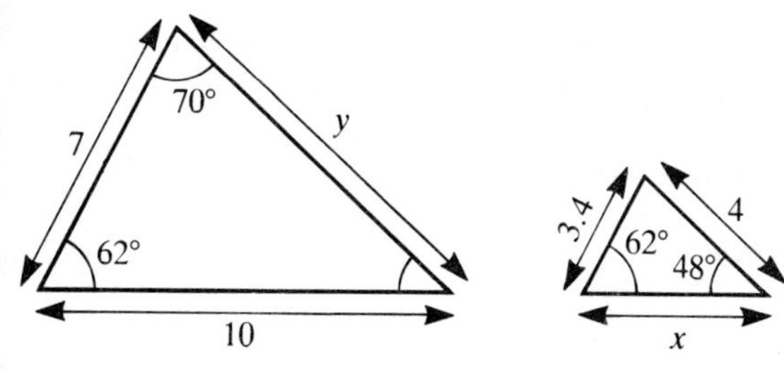

Figure 13.32 Diagram for problem 4.

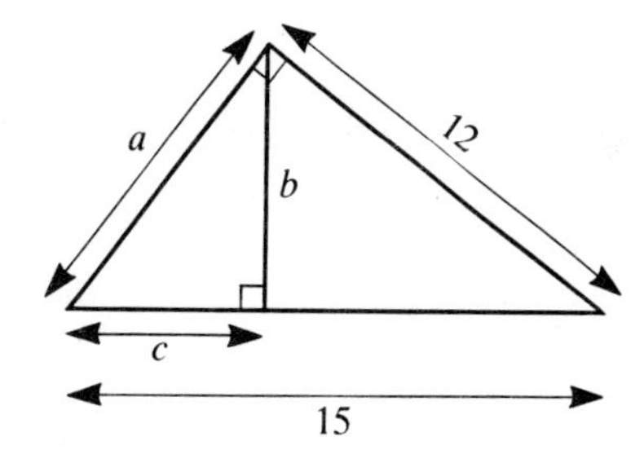

Figure 13.33 Diagram for problem 5.

Problems 13

1 Determine the angles marked x in Figure 13.29.

2 Determine the lengths marked x in the right-angled triangles shown in Figure 13.30.

3 State the four conditions for congruency of triangles. Determine whether or not the respective pairs of triangles drawn in Figure 13.31 are congruent, giving the reason.

4 Calculate the lengths x and y in the similar triangles of Figure 13.32.

5 Calculate the lengths a, b and c in the diagram of Figure 13.33.

6 Construct the following triangles using rule, protractor and compass as required.
(a) $\triangle$ABC where $\angle A = 65°$, $\angle C = 75°$, AB = 50 mm. Measure and record AC.
(b) $\triangle$ABC in which $\angle A = 100°$, $\angle B = \angle C$ and BC = 77 mm. Measure and record AB and AC. What type of triangle is $\triangle$ABC?
(c) $\triangle$ABC, given AB = 35 mm, BC = 50 mm, AC = 40 mm. Measure and record $\angle B$ and $\angle A$.

14 Geometric properties of circles

General learning objectives: to identify the geometric properties of circles.

14.1 Definitions of important terms relating to circles

A **circle** is a plane figure bounded by a curve traced out by a point moving at a constant distance from a fixed point. The fixed point is known as the **centre** of the circle. The boundary line of the circle, the periphery or perimeter of a circle, is known as the **circumference**, see Figure 14.1(a).

Any straight line joining the centre and any point on the circumference is known as a **radius**; any straight line passing through the centre and joining two 'diametrically' opposite points on the circumference is known as a **diameter**, see Figure 14.1(b). Note,

$$\text{diameter} = 2 \times \text{radius}, \quad d = 2r$$

The ratio of the circumference to the diameter for all circles, regardless of size, is a constant. This constant is denoted by the Greek letter π (pi). π is the first letter of the Greek word perimetros which means 'length of the perimeter'.

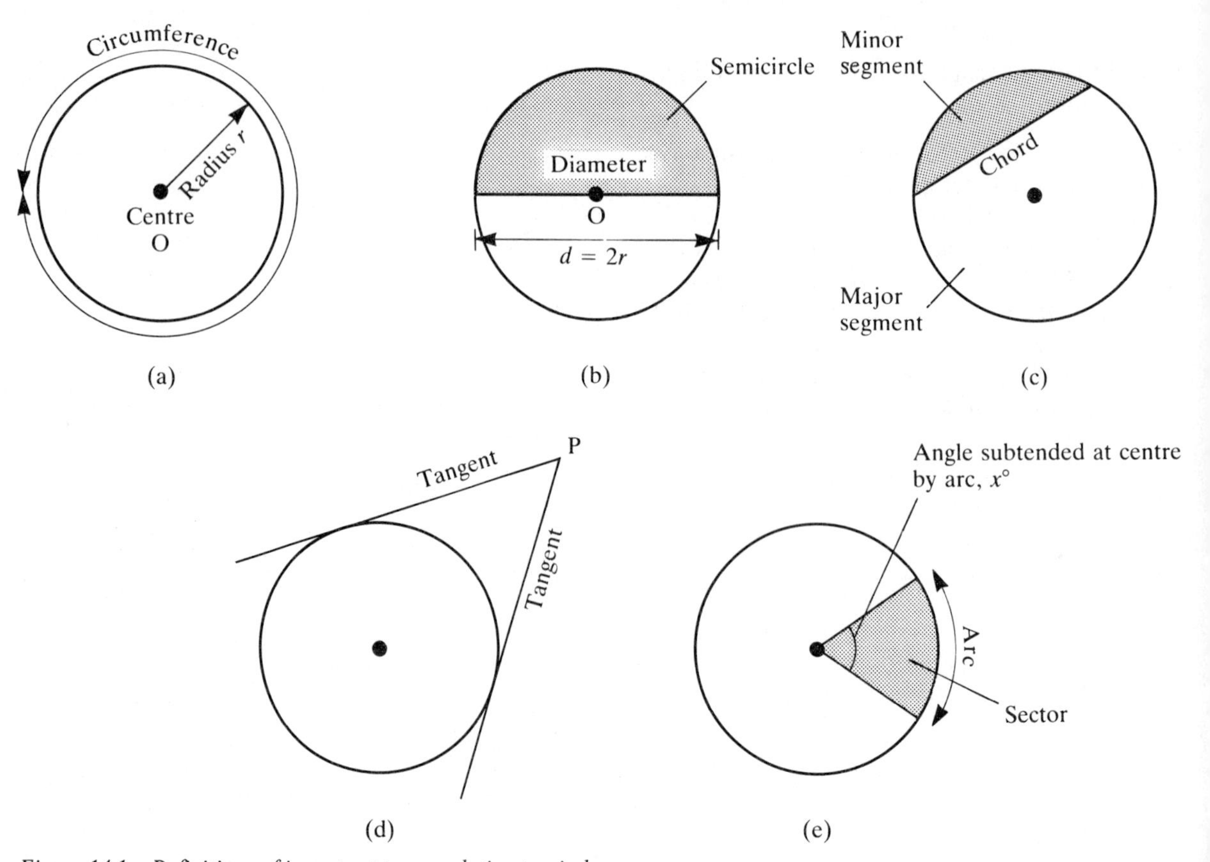

Figure 14.1 Definitions of important terms relating to circles.

$$\pi = \frac{\text{circumference of circle}}{\text{diameter of that circle}}$$

so if c = circumference
(perimeter or distance round circle)
d = diameter of circle $= 2r$, r = radius of circle

then as $\pi = \frac{c}{d}$,

$$c = \pi d = 2\pi r$$

π cannot be specified absolutely, it is a non-terminating, non recurring decimal, which has the value

$$\pi = 3.141\,592\,654 \text{ (to 9 decimal figure accuracy)}$$
$$= 3.142 \text{ (to 3 decimal places)}$$

The 'often used' approximation of $\pi = \frac{22}{7} = 3.1429$ is in fact only accurate to two decimal places.

The area of a circle (see also Chapter 12) is given by the formula

$$A = \pi r^2 = \frac{1}{4}\pi d^2$$

Any straight line joining two points on the circumference of a circle is known as a **chord**. A chord divides a circle into two **segments**, the larger one is known as the **major segment**, the smaller one as the **minor segment**, see Figure 14.1(c). A diameter is a special case of a chord which divides the circle into two equal parts, two **semicircles**, as shown in Figure 14.1(b).

A **tangent** to a circle is a straight line drawn from a point outside the circle which touches the circle at one point only. From an external point two tangents may be drawn to a circle, as shown in Figure 14.1(d). A **secant** is a straight line from the centre of a circle cutting the circumference and proceeding till it meets a tangent to the same circle.

An **arc** of a circle is any part of the circumference whose length is less than the circumference. A **sector** is a part of a circle bounded by an arc and two radii, as shown Figure 14.1(e). The angle **subtended** by an arc is the angle between the two radii joining the extreme points of the arc. If this angle is x°, then

$$\text{arc length} = c \times \frac{x^\circ}{360^\circ} = 2\pi r \times \frac{x^\circ}{360^\circ}$$

$$\text{sector area} = \pi r^2 \times \frac{x^\circ}{360^\circ}$$

14.2 Applications relating radius, diameter, circumference, etc. of circles

The important formulae for circles relating radius and diameter to circumference, area, arc length, sector area are:

circumference $c = \pi d = 2\pi r$

area $A = \frac{1}{4}\pi d^2 = \pi r^2$

diameter $d = 2r$, r = radius

arc length $= c \times \frac{\theta^\circ}{360} = 2\pi r \times \frac{\theta^\circ}{360} = \frac{\pi r \theta^\circ}{180}$

sector area $= \pi r^2 \times \frac{\theta^\circ}{360^\circ}$

where θ° = angle in degrees which arc and sector subtend at the circle centre.

These formulae are applied in the following examples.

Examples

1 Calculate the circumference and area of
(a) a circle diameter 100 mm
(b) a circle radius 0.25 mm
Take $\pi = 3.142$.

Solution

(a) Circumference $= \pi d = 3.142 \times 100$
$= 314.2$ mm *Ans*

Area $= \frac{1}{4}\pi d^2 = 0.25 \times 3.142 \times 100^2$
$= 7855 \text{ mm}^2$ *Ans*

(b) Circumference $= 2\pi r = 2 \times 3.142 \times 0.25$
$= 1.571$ mm *Ans*

Area $= \pi r^2 = 3.142 \times 0.25^2$
$= 0.1964 \text{ mm}^2$ *Ans*

2 A 2.4 metre length of wire is bent into the shape of a circle. Calculate the radius and area of the resulting circle. Take $\pi = 3.142$.

Solution

Let radius of resulting circle be r, then as the circumference will equal the wire length, we have

$$2\pi r = 2.4$$

$$r = \frac{2.4}{2\pi} = \frac{2.4}{2 \times 3.142} = 0.382 \text{ m} \quad \textit{Ans}$$

and area of resulting circle

$$A = \pi r^2 = 3.142 \times 0.382^2 = 0.458\,\text{m}^2 \quad \textit{Ans}$$

3 The diameter of the wheels of a cycle is 0.71 m. Calculate the distance travelled after 500 complete revolutions of the wheels. Take $\pi = 3.142$.

Solution

Distance travelled in one revolution
= circumference of wheel
$= \pi d = 3.142 \times 0.71 = 2.231$ m,
so total distance travelled in 500 revolutions,

$$500 \times \pi d = 500 \times 2.231 = 1115.5\,\text{m} \quad \textit{Ans}$$

4 Calculate the length of the arcs marked s and sector areas A in the circles of Figure 14.2. Take $\pi = 3.142$.

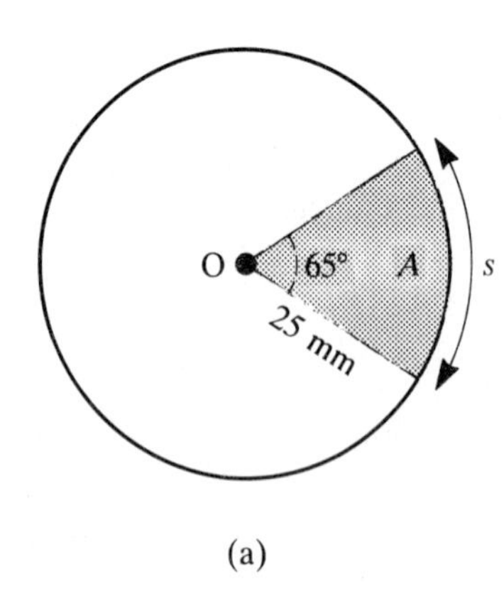

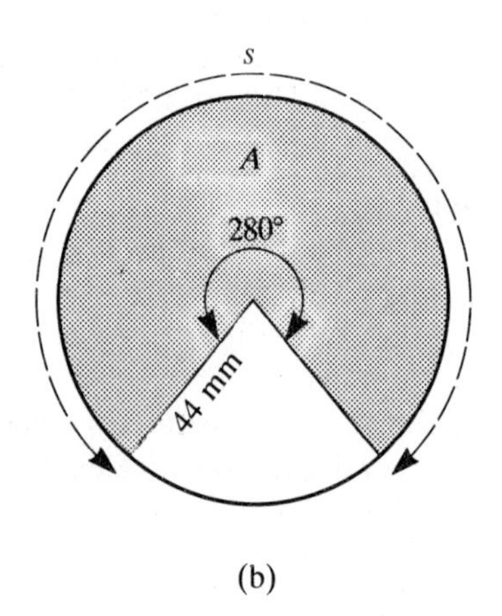

Figure 14.2 Diagrams for example 4.

Solution

(a) arc length $s = 2\pi r \times \dfrac{\theta^\circ}{360^\circ}$

where $\theta = 65^\circ$ and $r = 25$ mm

i.e. $s = 2 \times 3.142 \times 25 \times \dfrac{65}{360}$

$= 28.37$ mm *Ans*

Area of sector, $A = \pi r^2 \times \dfrac{\theta}{360}$

$= 3.142 \times 25^2 \times \dfrac{65}{360}$

$= 354.6\,\text{mm}^2$ *Ans*

(b) For this case $r = 44$ mm and θ is the reflex angle 280°, so

arc length $s = 2\pi r \times \dfrac{\theta}{360}$

$= 2 \times 3.142 \times 44 \times \dfrac{280}{360}$

$= 215.1$ mm *Ans*

area $A = \pi r^2 \times \dfrac{\theta}{360} = 3.142 \times 44^2 \times \dfrac{280}{360}$

$= 4731\,\text{mm}^2$ *Ans*

5 Geostationary satellites orbit the earth in a circular orbit at a distance of approximately 42 000 km (26 000 miles) from the centre of the earth. They take 24 hours to travel one complete orbit and thus have the important property that they keep in step with the earth's own rotation so that they appear stationary in space.

Calculate the speed that the geostationary satellites travel through space in kilometres per hour.

Solution

The distance travelled in one complete orbit

= circumference of circle of radius 42 000 km
$= 2\pi r = 2 \times 3.142 \times 42\,000 = 263\,928$ km

and this takes the time of 24 hours, so

speed of satellite $= \dfrac{\text{distance travelled}}{\text{time}}$

$= \dfrac{263\,928\,\text{km}}{24\,\text{h}}$

$= 10\,997$ km/h
$\approx 11\,000$ km/h *Ans*

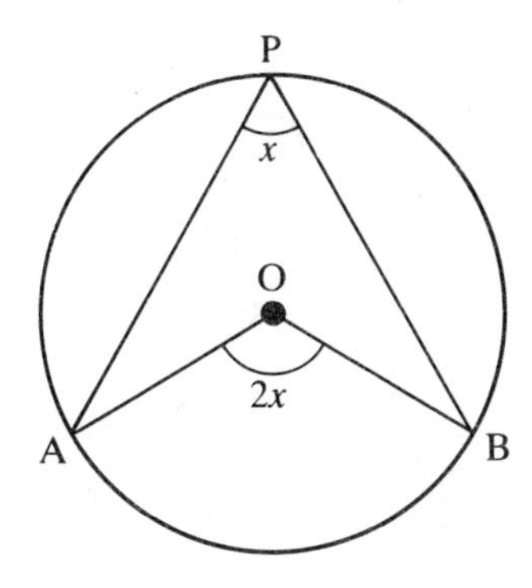

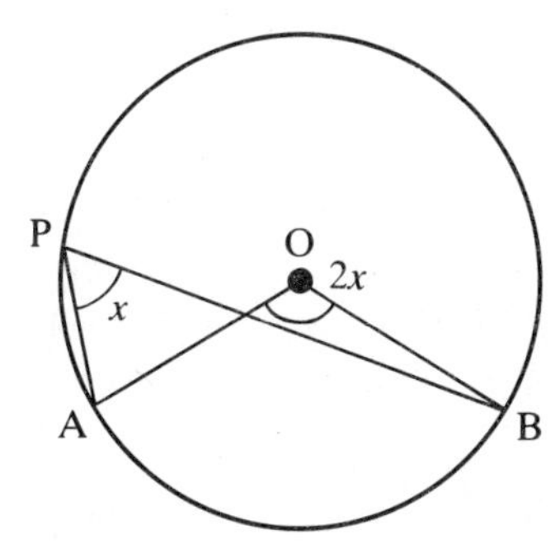

Figure 14.3 Angle subtended at centre = 2 × angle subtended at the circumference.

14.3 Angle relationships in a circle

Some important angle relationships and properties for angles in a circle are listed below.

1 **The angle subtended at the centre of a circle is twice the angle subtended at the circumference.** This theorem is illustrated in Figure 14.3. $\angle$AOB is the angle subtended at the circle centre O by chord AB, P is any point on the arc (part of circumference) between A and B

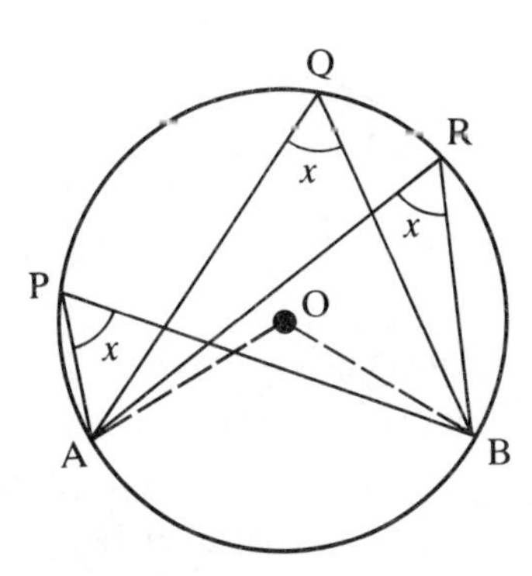

Figure 14.4 'Angles on the same arc are equal'.

and so $\angle$APB is the angle subtended on the circumference.

2 **Angles subtended at the circumference by the same chord are equal.**
It follows from 1 that all angles subtended by the same chord are equal or briefly 'angles on the same arc are equal'. This property is illustrated in Figure 14.4:

$$\angle P = \angle Q = \angle R = \frac{1}{2} \times \angle AOB$$

3 **The angle in a semicircle is equal to 90°.**
It also follows from 1 that an angle in a semicircle (the angle subtended by a diameter) is 90°, since the angle subtended by the diameter at the centre is obviously 180° and the angle subtended at the circumference (in this case a semicircle portion) is half this, i.e. $\frac{1}{2} \times 180° = 90°$, see Figure 14.5.

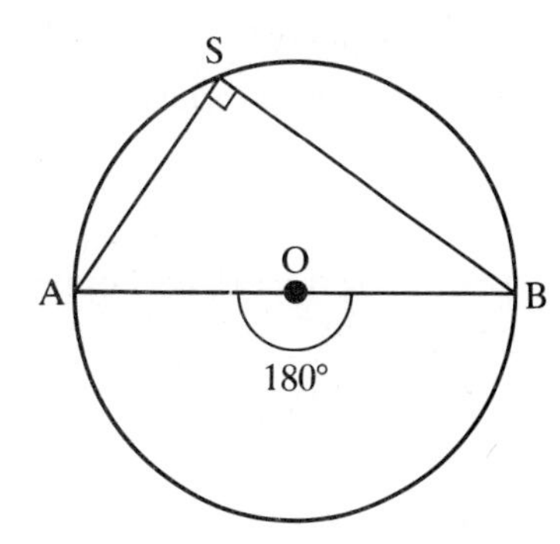

Figure 14.5 Angles contained in a semi-circle are right angles.

4 **The angle between a radius and a tangent meeting at the same point on a circle is a right angle.** This property is illustrated in Figure 14.6. Tangent PT and radius OT are perpendicular to each other, $\angle$PTO = 90°.

5 **Angle properties of a cyclic quadrilateral.**
If a circle can be drawn through the four corners of a quadrilateral, then the quadrilateral is known as a **cyclic quadrilateral.** A cyclic quadrilateral possesses the following two angle properties:

(i) the internal opposite angles add up to 180°, i.e. in Figure 14.7

$$\angle A + \angle C = 180°$$
$$\angle B + \angle D = 180°$$

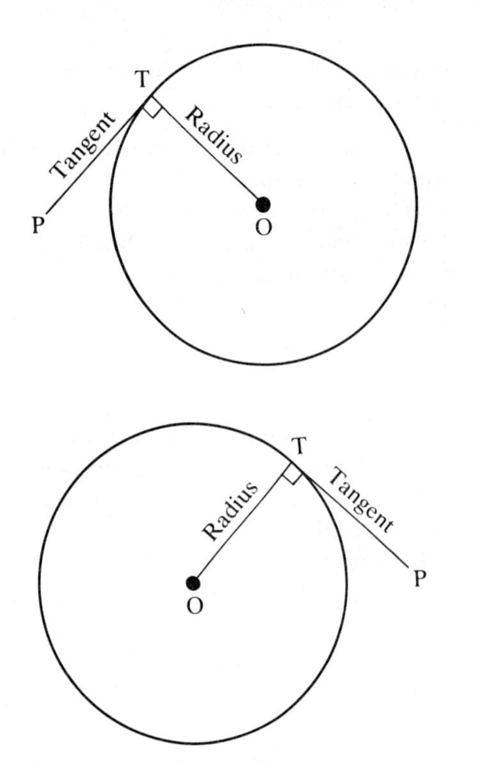

Figure 14.6 Angle between a radius and a tangent is a right angle.

(ii) the interior angle = external opposite angle, e.g. in Figure 14.7

$\angle A = \angle BCX$

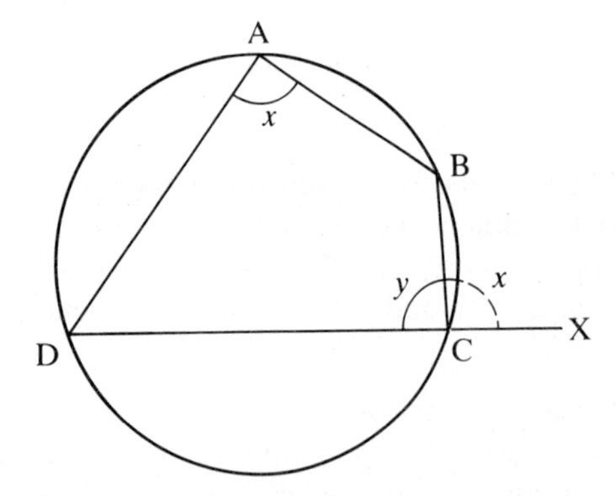

Figure 14.7 Cyclic quadrilateral angle properties: internal opposite angles add to 180° (are supplimentary), internal angle = external opposite angle.

Examples

1 Determine the values of the angles marked x and y in the circles of Figure 14.8.

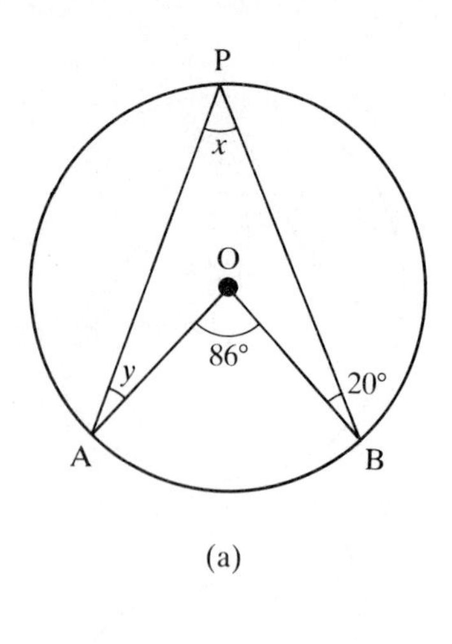

(a)

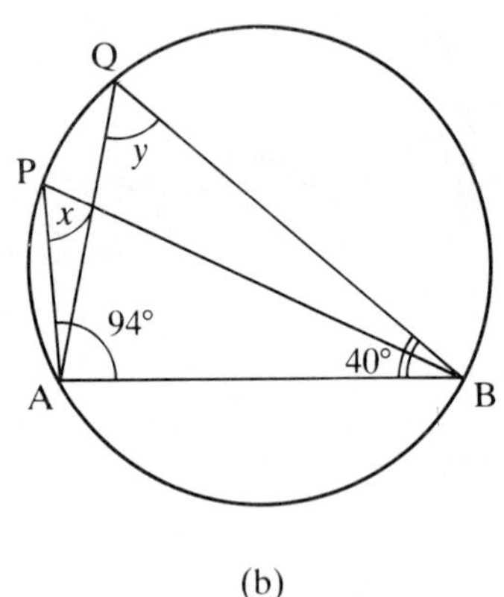

(b)

Figure 14.8 Diagrams for example 1.

Solution

(a) Since angle subtended at the centre, 86°, is twice that subtended at the circumference,

$2x = 86°, \quad x = 43° \quad$ *Ans*

If we join PO, then as $\triangle$ POB is isosceles (OP = OB, radii)

$\angle OPB = \angle OBP = 20°$

hence $\angle APO = \angle P - \angle OPB$

$= 43° - 20° = 23°$

but $\triangle$ APO is also isosceles (OA = OP, radii) and so

$\angle PAO = y = \angle APO = 23°$,

i.e. $y = 23°$ *Ans*

(b) Using the fact that the sum of the internal angles in a triangle equal 180°, we have for $\triangle$ APB,

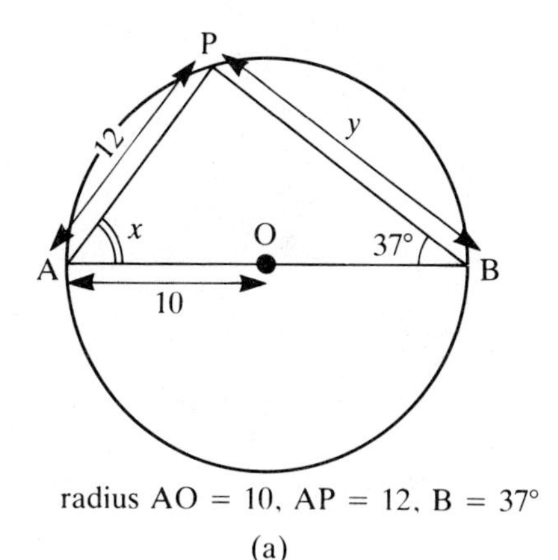

radius AO = 10, AP = 12, B = 37°

(a)

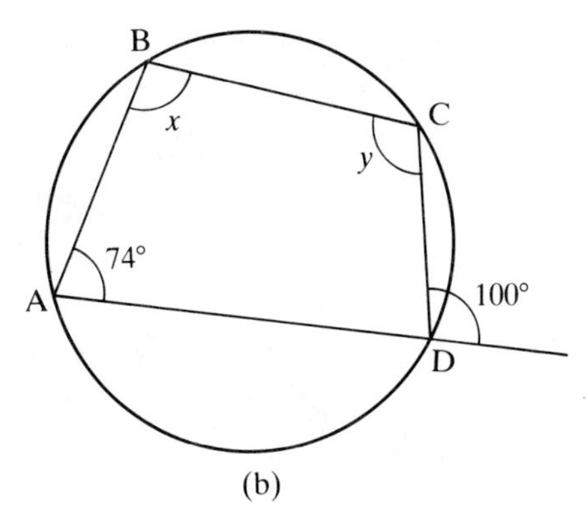

(b)

Figure 14.9 Diagrams for example 2.

$$x + 40 + 94 = 180°$$
so $x = 180 - 40 - 94 = 46°$ *Ans*

Also, as angles subtended at the circumference by the same chord are equal, we have

$$\angle P = x = 46° = \angle Q = y,$$
i.e. $y = 46°$ *Ans*

2 Determine the unknown quantities x and y in the diagrams of Figure 14.9.

Solution

(a) Since the angle in a semi-circle is 90°,

$$\angle P = 90° \quad \text{in Figure 14.9(a)}$$

and as sum of internal angles in $\triangle$APB is 180°,

$$\angle P + x + 37° = 180°$$
$$x = 180° - 90 - 37$$
$$= 90 - 37 = 53° \quad \textit{Ans}$$

$\triangle$APB is a right-angled triangle, and applying Pythagoras' theorem,

$$AB^2 = AP^2 + PB^2$$
$$20^2 = 12^2 + y^2$$
so $y^2 = 20^2 - 12^2 = 400 - 144 = 256$

$y = \sqrt{256} = 16$ *Ans*

(b) Quadrilateral ABCD in Figure 14.9(b) is cyclic, hence internal angle

$$\angle B = x = \text{external opposite angle}$$
$$= 100°, \text{ i.e. } x = 100° \quad \textit{Ans}$$
$$\angle A + \angle C = 180°$$
(internal opposite angles add to 180°)

so $74° + y = 180°$,

$y = 180 - 74 = 106°$ *Ans*

Test and problems 14

Multiple choice test: MT 14

Answer block:

Question No.	0	1	2	3	4	5	6	7	8
Answer	c								

Enter your answer, that is a, b, c or d in the column under the question number in the answer block above.
Note that question Qu. 0 has already been worked out and the answer inserted.

Qu. 0 In Figure 14.10 tangent TP touches the circle at P, OT = 20 and the radius OP = 10. Calculate length PT.
Ans (a) 10 (b) 15.24 (c) 17.32 (d) 18.0

Solution

Since the tangent and radius at the point at which the tangent touches the circle are at right angles, $\angle P = 90°$ and $\triangle$OPT is a right-angled triangle. Hence, applying Pythagoras' theorem,

$$OT^2 = OP^2 + PT^2$$
$$20^2 = 10^2 + PT^2$$
so $PT^2 = 20^2 - 10^2 = 400 - 100 = 300$

$PT = \sqrt{300} = 17.32$ *Ans*

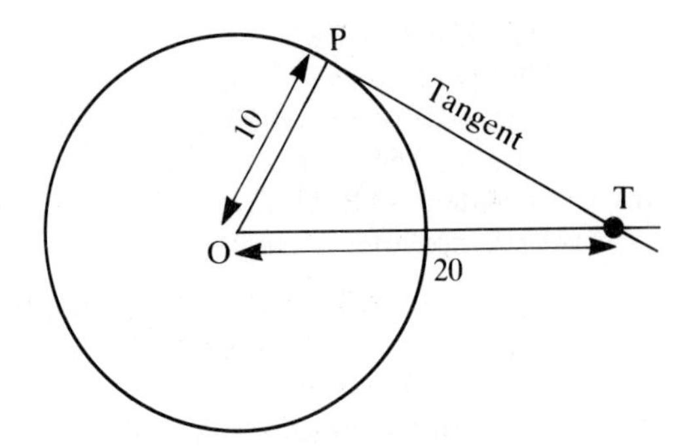

Figure 14.10 Diagram for Qu 0.

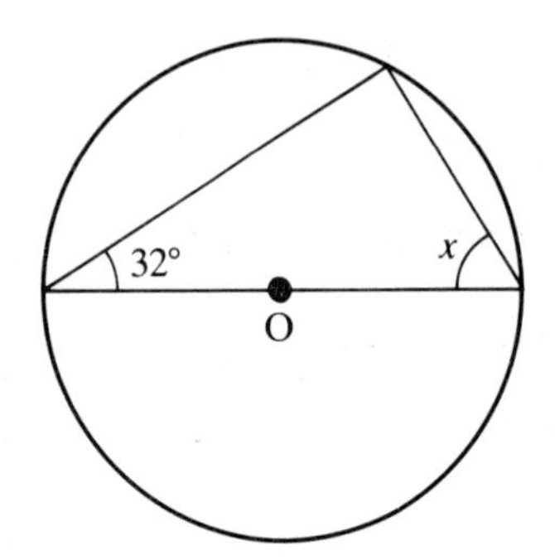

Figure 14.11 Diagram for Qu 1.

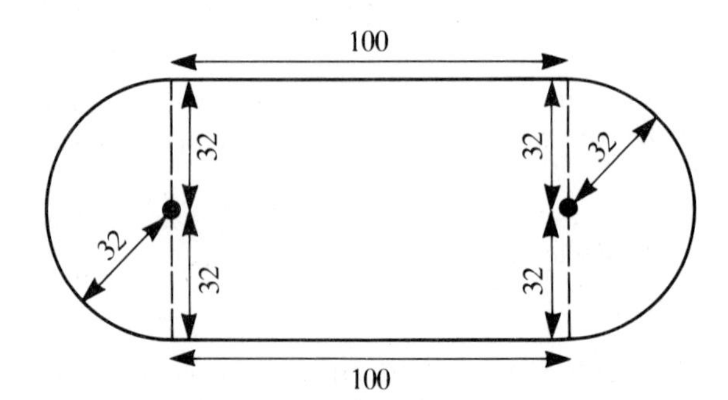

Figure 14.12 Diagram for Qu 2.

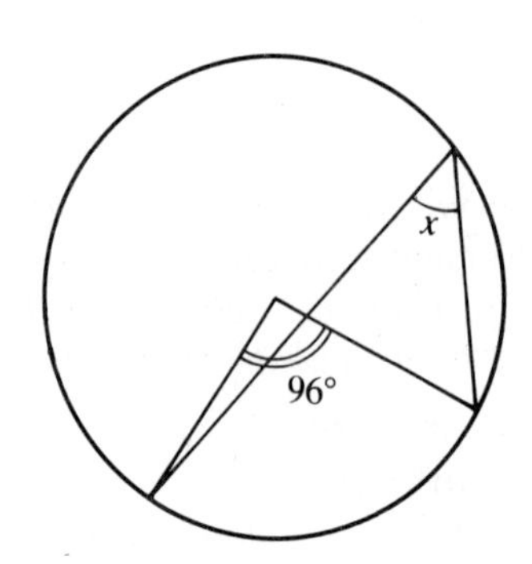

Figure 14.13 Diagram for Qu 3.

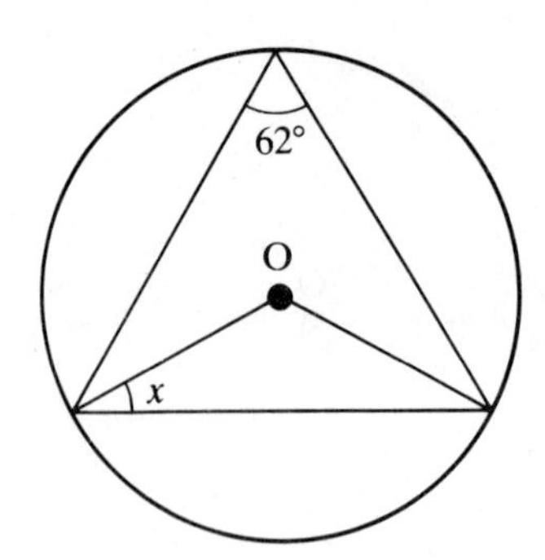

Figure 14.14 Diagram for Qu 4.

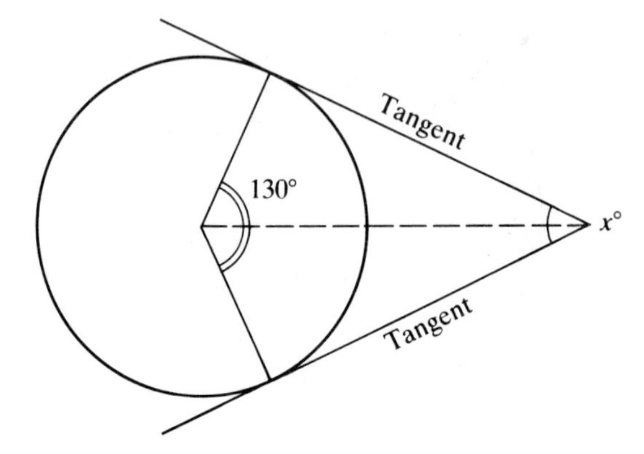

Figure 14.15 Diagram for Qu 5.

Hence (c) is the correct answer and therefore 'c' is entered in the answer block under Qu. 0 as shown.

Now carry on with the test; take $\pi = 3.142$ where needed.

Qu. 1 Determine the angle x in the diagram of Figure 14.11.
Ans (a) 32° (b) 58° (c) 64° (d) 78°

Qu. 2 Determine the total distance, total periphery, of the diagram shown in Figure 14.12.
Ans (a) 328 (b) 502 (c) 450 (d) 401

Qu. 3 Determine the angle x in the diagram of Figure 14.13.
Ans (a) 96° (b) 48° (c) 54° (d) 68°

Qu. 4 Determine the angle x in the diagram of Figure 14.14.
Ans (a) 31° (b) 28° (c) 45° (d) 62°

Qu. 5 Determine the angle x in the diagram of Figure 14.15.
Ans (a) 65° (b) 25° (c) 50° (d) 62.4°

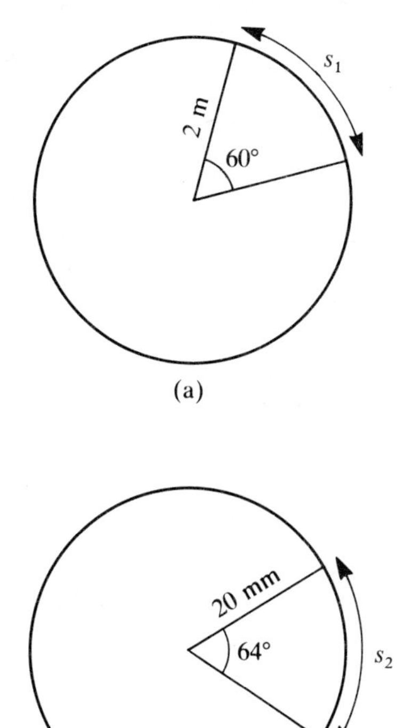

Figure 14.16 Diagrams for problem 4.

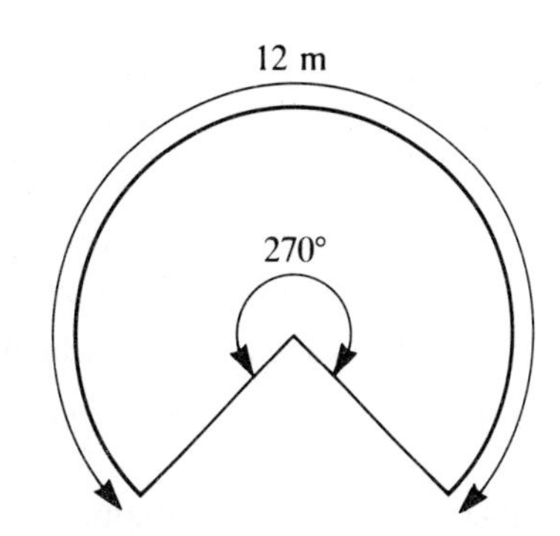

Figure 14.17 Diagrams for problem 5.

Qu. 6 Which of the following circles has an area of equal magnitude to its circumference? The circle with a radius of

Ans (a) $r = \pi$ (b) $r = 1$ (c) $r = 10$ (d) $r = 2$

Qu. 7 The second hand of a clock is 8 cm in length. Estimate the distance travelled by the tip of this hand recording the seconds in one week.

Ans (a) 422 m (b) 507 m (c) 5067 m (d) 20.3 km

Qu. 8 Construct a right-angled triangle whose hypotenuse is 60 mm and the length of a

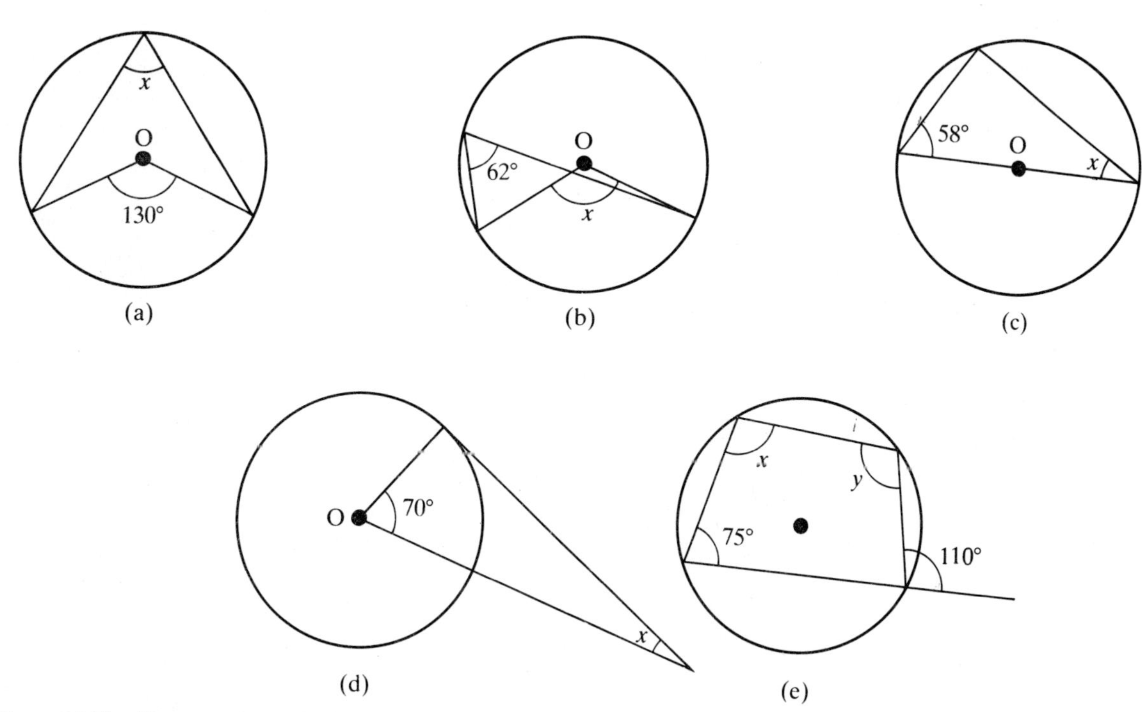

Figure 14.18 Diagrams for problem 6.

second side is 25 mm. Measure the smallest angle in this triangle. Its value is
Ans (a) 15° (b) 43° (c) 37° (d) 25°

Problems 14 (Take $\pi = 3.142$ where needed)

1 Calculate the circumference of the following circles:
(a) diameter 50 mm (b) diameter 42.64 m
(c) radius 6374 km

2 Calculate the distance moved in 1 second by a point on the periphery of a grinding wheel of diameter 112 mm when rotating at the speed of 3000 revolutions per minute.

3 The diameter of the earth is approximately 12 750 km. The earth spins on its axis making one revolution in 24 hours. Determine the speed of rotation for a point on the earth's equator in kilometres per hour.

4 Calculate the arc lengths s_1 and s_2 in the diagrams of Figure 14.16.

5 The arc length of the partial circle shown in Figure 14.17 is 12 m and the angle subtended at the centre is 270°. Determine the radius of the circle.

6 Determine the unknown angles in the diagrams of Figure 14.18.

15 Introduction to trigonometry

General learning objectives: to solve right-angled triangles for angles and lengths using sine, cosine and tangent functions.

15.1 Introduction to trigonometry and sine, cosine and tangent functions

Trigonometry is a branch of mathematics which is defined as 'the science of determining the sides and angles of triangles by means of certain parts that are given'. In particular, trigonometry enables us to calculate the lengths and angles of triangles and other geometric figures assisted by the trigonometric ratios or functions of sine, cosine and tangent. Trigonometry has wide applications and is of high importance in astronomy, surveying, navigation, science and engineering.

Relationships between the sides and angles of right-angled triangles are expressed in terms of the trigonometric functions of *sine, cosine and tangent* of its angles which, with reference to Figure 15.1, are defined for the angle θ as:

sine of the angle θ, $\sin\theta = \dfrac{\text{Perpendicular}}{\text{Hypotenuse}} = \dfrac{P}{H}$

cosine of the angle θ, $\cos\theta = \dfrac{\text{Base}}{\text{Hypotenuse}} = \dfrac{B}{H}$

tangent of the angle θ, $\tan\theta = \dfrac{\text{Perpendicular}}{\text{Base}} = \dfrac{P}{B}$

Note: function names are normally always abbreviated to sin, cos and tan.

The definitions of sin, cos and tan can be remembered by memorizing the 'phrase':

Percy Has Been Here Pinching Buns

$$\frac{P}{H} = \sin\theta \quad \frac{B}{H} = \cos\theta \quad \frac{P}{B} = \tan\theta$$

where P = Perpendicular length or side *opposite* the angle
B = Base, the side *adjacent* to the angle
H = Hypotenuse, the side oposite the right-angle

We may also remember the definition of sin, cos and tan in terms of the sides *opposite, adjacent to* the angle and the *hypotenuse* as, see Figure 15.2:

$$\sin\theta = \frac{\text{opposite}}{\text{hypotenuse}}$$

$$\cos\theta = \frac{\text{adjacent}}{\text{hypotenuse}}$$

$$\tan\theta = \frac{\text{opposite}}{\text{adjacent}}$$

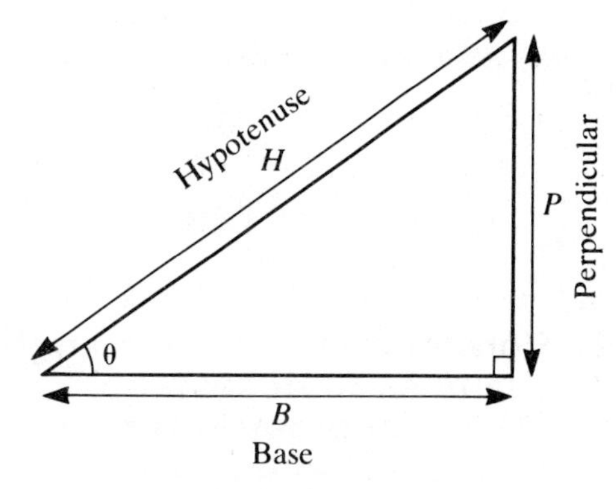

Figure 15.1 Definitions for trigonometric functions::

$\sin\theta = \dfrac{P}{H}$, $\cos\theta = \dfrac{B}{H}$, $\tan\theta = \dfrac{O}{A}$ or $\dfrac{P}{B}$

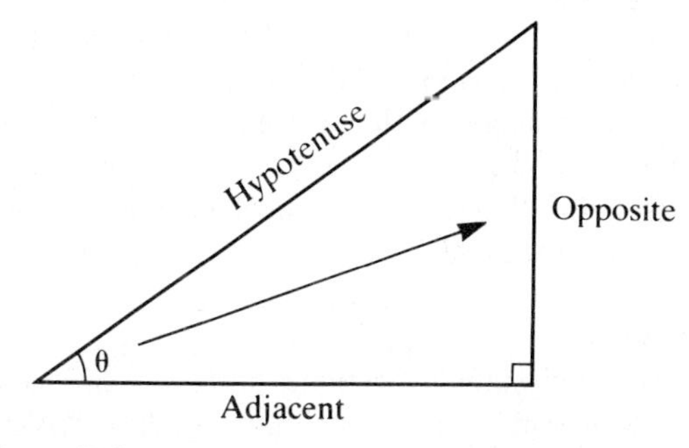

Figure 15.2.

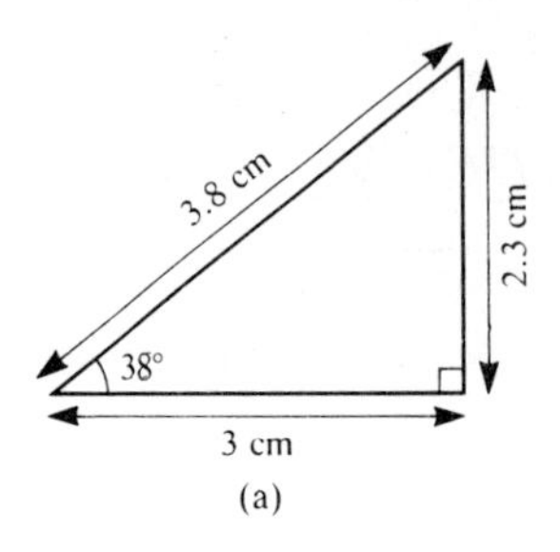

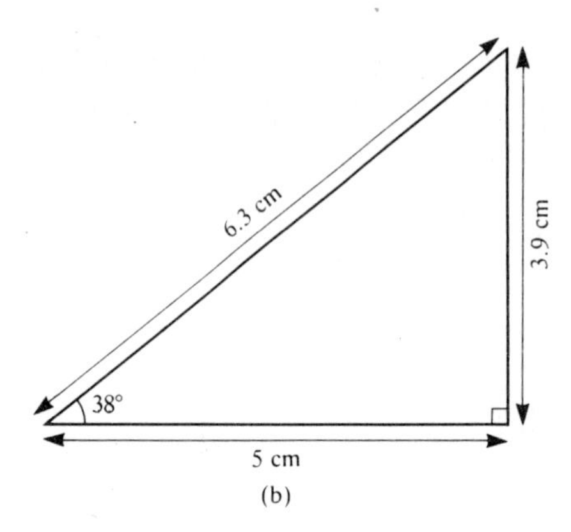

Figure 15.3 Right-angled triangles constructed for example 1.

15.2 Construction of right-angled triangles to determine: (1) trigonometric function values, (2) angles from knowledge of trigonometric function values

Examples

1 Using a ruler and protractor construct a right-angle with one angle of 38° and base (adjacent) equal to 3 cm. Use this triangle to find the values of sin 38°, cos 38° and tan 38°. Construct a *similar* triangle with base equal to 5 cm and show, within the accuracy limits of your construction, that the trigonometric functions for 38° are identical.

Solution

The right-angled triangle with $\theta = 38°$ and base = 3 cm is drawn in Figure 15.3(a). Taking measurements from the figure:

perpendicular or opposite = 2.3 cm;
hypotenuse = 3.8 cm

so $\sin 38° = \dfrac{P}{H} = \dfrac{2.3}{3.8} = 0.61$

$\cos 38° = \dfrac{B}{H} = \dfrac{3}{3.8} = 0.79$

$\tan 38° = \dfrac{P}{B} = \dfrac{2.3}{3} = 0.77$

The right-angled triangle with $\theta = 38°$ but with base = 5 cm is drawn in Figure 15.3(b) and from measurement,

perpendicular = 3.9; hypotenuse = 6.3

so $\sin 38° = \dfrac{3.9}{6.3} = 0.62, \quad \cos 38° = \dfrac{5}{6.3} = 0.79$

$\tan 38° = \dfrac{3.9}{5} = 0.78$

so within the limits of our construction accuracy the trigonometric functions are equal – exactly as we should expect.

2 By constructing appropriate right-angled triangles, find the angles which correspond to (a) $\tan\theta = 2$, (b) $\cos\theta = \frac{1}{2}$, (c) $\sin\theta = 0.8$.

Solution

(a) $\tan\theta = \dfrac{P}{B} = 2$

so to find the angle θ construct a right-angled triangle with perpendicular height $P = 2$ units and base $B = 1$ unit as shown in Figure 15.4(a). On measuring the angle θ with a protractor we obtain

$\theta = 63°$ *Ans*

(b) $\cos\theta = \dfrac{B}{H} = \dfrac{1}{2}$

so in this case to determine θ we construct a right-angled triangle with base $B = 1$ unit and hypotenuse $H = 2$ units. To do this we first draw in the base of 1 unit length and construct a right-angle at one end of this line. With a compass radius $H = 2$ units and centred at the other end of the base draw an arc to intersect the perpendicular. Draw in the line from this point of intersection to the base point used to construct the arc, i.e.

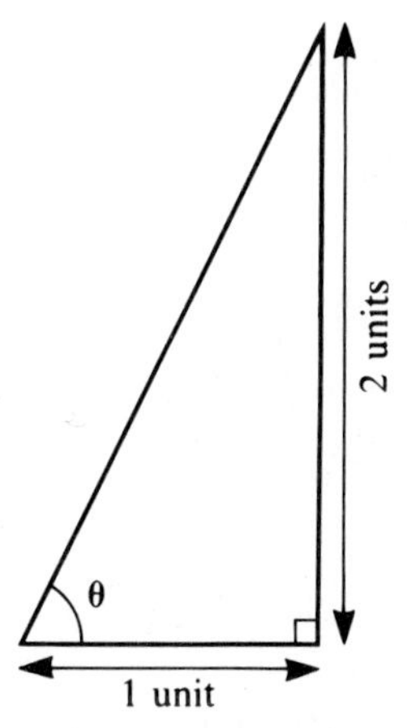

(a) Construction of right-angled triangle with perpendicular $P = 2$ and base $B = 1$ to determine θ when $\tan \theta = 2$

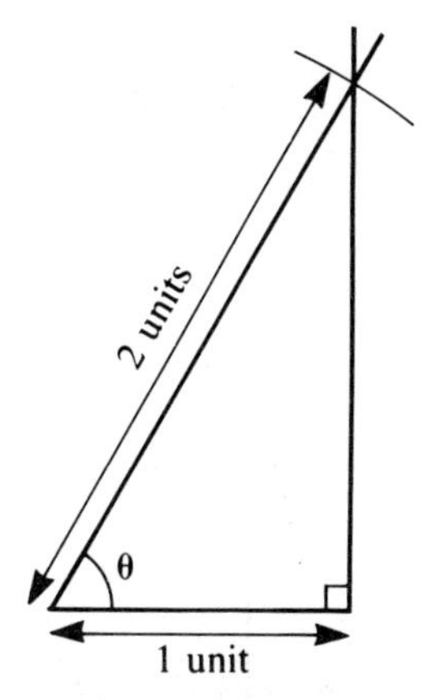

(b) Construction of right-angled triangle with base $B = 1$ and hypotenuse $H = 2$ to determine θ when $\cos \theta = \frac{1}{2}$

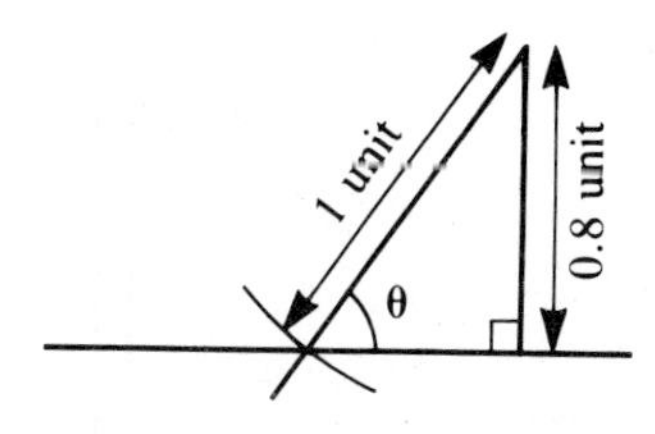

(c) Construction of right-angled triangle with perpendicular $P = 0.8$ and hypotenuse $H = 1$ to determine θ when $\sin \theta = 0.8$

Figure 15.4 Construction for example 2.

draw in the hypotenuse. The construction is shown in Figure 15.4(b). Mark in θ and measure with protractor:

$$\theta = 60° \quad Ans$$

(c) $\sin \theta = \dfrac{P}{H} = 0.8$

To find θ we must draw a right-angled triangle with perpendicular $P = 0.8$ units and hypotenuse $H = 1$ unit. First construct a base line and the perpendicular of length 0.8 units. Then with compass centred at the top end of the perpendicular draw an arc of length $H = 1$ unit to cut the base, see Figure 15.4(c). From this diagram,

$$\theta = 53° \quad Ans$$

15.3 Use of tables and electronic calculators to determine the values of trigonometric functions

As previously considered in Chapter 4, section 4.5, sine, cosine and tangent values for angles in the range 0° to 90° are listed in most mathematical tables booklets under *natural* sines, cosines and tangents. See section 4.5 for the use of these tables.

In Chapter 5, section 5.3, the use of an electronic calculator for determining both the trigonometric functions given the angle and also for determining the angle given the function value is explained. See section 5.3 for the use of the sine, cosine and tangent keys and also the inverse or $\sin^{-1}$, $\cos^{-1}$, $\tan^{-1}$ operations.

Examples

1 Use 4-figure tables and/or an electronic calculator to determine (a) $\sin 38° 44'$ (b) $\cos 72° 15'$ (c) $\tan 67.8°$

Answers

(a) $\sin 38° 44' = 0.6257$ *Ans*

Note: tables normally list trigonometric functions in terms of degrees (°) and minutes ('):

$$1° = 60'$$

Calculators require the angle in degrees so the 44′ must first be converted into a decimal fraction:

$$44' = 44/60 = 0.7333^\circ$$

and hence 38° 44′ must be entered as 38.7333°

(b) $\cos 72^\circ 15' = 0.3049$ *Ans*
(c) $\tan 67.8^\circ = 2.4504$ *Ans*

2 Determine the angles for which

(a) $\tan\theta = 3.2361$ (b) $\cos\theta = 0.9294$

Answers

(a) Using the inverse (INV) and TAN keys or $\tan^{-1}$ key:

$\theta = \tan^{-1} 3.2361 = 72.83^\circ$ *Ans*

or as $0.83^\circ = 0.83 \times 60' = 50'$

$\theta = 72^\circ 50'$ *Ans*

(b) $\theta = \cos^{-1} 0.9294$
$= 21.66^\circ$ *Ans*
$= 21^\circ 40'$ *Ans*

15.4 Trigonometric function values for standard triangles: 30°, 60°, 90° and 45°, 45°, 90° triangles

The trigonometric function values for 30° and 60°, and for 45° can be found directly from the 'standard' right angles which contain these angles.

The sides for a 30°, 60°, 90° triangle, see Figure 15.5(a) are in the ratio $1{:}\sqrt{3}{:}2$ and so on applying the definitions for sine, cosine and tangent, we obtain:

$$\sin 30^\circ = \frac{\text{opposite}}{\text{hypotenuse}} = \frac{1}{2}$$

$$\cos 30^\circ = \frac{\text{adjacent}}{\text{hypotenuse}} = \frac{\sqrt{3}}{2}$$

$$\tan 30^\circ = \frac{\text{opposite}}{\text{adjacent}} = \frac{1}{\sqrt{3}}$$

since for 30° opposite = 1, adjacent = $\sqrt{3}$, and hypotenuse = 2.

For 60°: opposite = $\sqrt{3}$ and adjacent = 1, so

$$\sin 60^\circ = \frac{\sqrt{3}}{2}$$

$$\cos 60^\circ = \frac{1}{2}$$

$$\tan 60^\circ = \sqrt{3}$$

The sides for a 45°, 45°, 90° triangle, see Figure 15.5(b), are in the ratio $1{:}1{:}\sqrt{2}$, so

$$\sin 45^\circ = \frac{1}{\sqrt{2}} = \frac{1}{\sqrt{2}} \times \frac{\sqrt{2}}{\sqrt{2}} = \frac{\sqrt{2}}{2}$$

$$\cos 45^\circ = \frac{1}{\sqrt{2}} = \frac{\sqrt{2}}{2}$$

$$\tan 45^\circ = \frac{1}{1} = 1$$

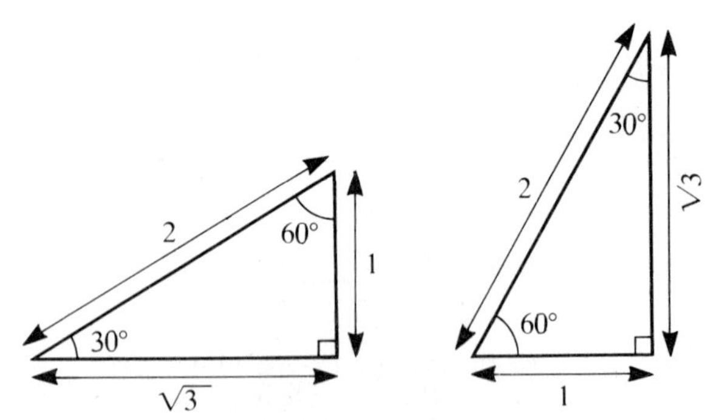

(a) 30°, 60°, 90° triangle : sides in ratio 1 : √3 : 2

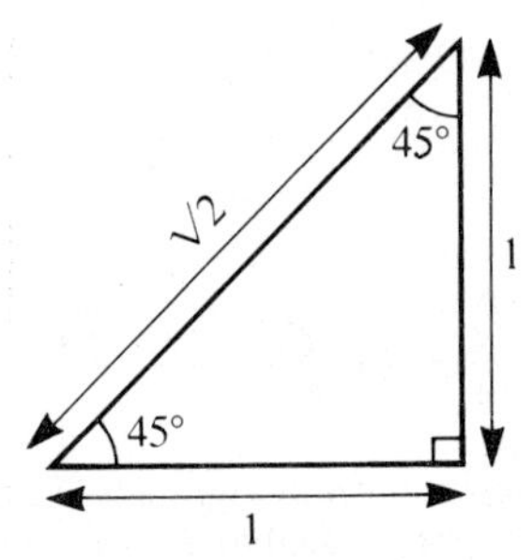

(b) 45°, 45°, 90° triangle : sides in ratio 1 : 1 √2

Figure 15.5 Standard triangles and direct deduction of trigonometric function ratios for 30°, 60°, and 45°.

15.5 Some important trigonometric relationships:

The following three relationships hold for the trigonometric functions:

1 $\cos\theta = \sin(90-\theta), \quad \sin\theta = \cos(90-\theta)$

2 $\tan\theta = \dfrac{\sin\theta}{\cos\theta}$

3 $\sin^2\theta + \cos^2\theta = 1.$

These results can easily be proved. On applying the definition for sine and cosine in the right-angled triangle of Figure 15.6 for the angles $\angle B = \theta$ and $\angle A = 90-\theta$, we have

$$\sin\theta = \frac{AC}{H}, \quad \cos\theta = \frac{BC}{H}$$

and $\cos(90-\theta) = \dfrac{AC}{H}, \quad \sin(90-\theta) = \dfrac{BC}{H},$

hence $\sin\theta = \cos(90-\theta)$ and $\cos\theta = \sin(90-\theta)$
Furthermore,

$$\frac{\sin\theta}{\cos\theta} = \frac{AC}{H} \div \frac{BC}{H} = \frac{AC}{BC}$$

but $\tan\theta = \dfrac{AC}{BC}$

so $\tan\theta = \dfrac{\sin\theta}{\cos\theta}$

Finally, on applying Pythagoras' theorem to the triangle of Figure 15.6, we have

$$AC^2 + BC^2 = H^2$$

but $AC = H\sin\theta$ and $BC = H\cos\theta$
so $H^2\sin^2\theta + H^2\cos^2\theta = H^2$

and on dividing throughout by H^2, we obtain

$$\sin^2\theta + \cos^2\theta = 1$$

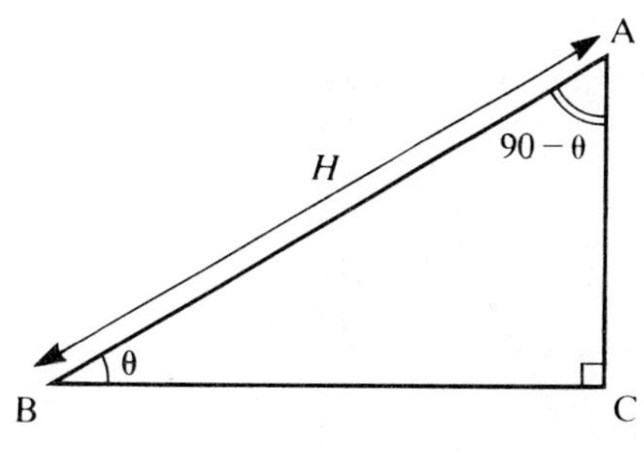

Figure 15.6.

Examples

1 (a) If $\cos 54° = 0.5878$, find $\sin 36°$
(b) If $\sin 79° = 0.9816$, find $\cos 11°$
(c) If $\sin 15° = 0.2588$, $\cos 15° = 0.9659$, determine $\sin 75°$, $\cos 75°$, $\tan 15°$, $\tan 75°$

Solution

(a) $\cos 54° = \sin(90-54) = \sin 36°$
hence $\sin 36° = 0.5878$ *Ans*

(b) $\sin 79° = \cos(90-79) = \cos 11° = 0.9816$ *Ans*

(c) $\sin 15° = \cos(90-15) = \cos 75° = 0.2588$ *Ans*

$\cos 15° = \sin(90-15) = \sin 75° = 0.9659$ *Ans*

also $\tan 15° = \dfrac{\sin 15}{\cos 15} = \dfrac{0.2588}{0.9659} = 0.2679$ *Ans*

and $\tan 75° = \dfrac{\sin 75}{\cos 75} = \dfrac{0.9659}{0.2588} = 3.732$ *Ans*

2 Using the identities

$$\sin^2\theta + \cos^2\theta = 1 \text{ and } \tan\theta = \sin\theta/\cos\theta$$

determine $\cos 25°$ and $\tan 25°$, given $\sin 25° = 0.4226$

Solution

Since $\sin^2\theta + \cos^2\theta = 1$

$\cos^2\theta = 1 - \sin^2\theta$

and $\cos\theta = \sqrt{(1-\sin^2\theta)}$

so $\cos 25° = \sqrt{(1-0.4226^2)}$
$= \sqrt{(1-0.1786)}$
$= \sqrt{0.8214}$
$= 0.9063$ *Ans*

Also as $\tan\theta = \sin\theta/\cos\theta$

$\tan 25° = \sin 25/\cos 25 = 0.4226/0.9063$
$= 0.4663$ *Ans*

15.6 Applications of trigonometry to practical problems

Examples

1 Determine the lengths x and y in the triangles of Figure 15.7.

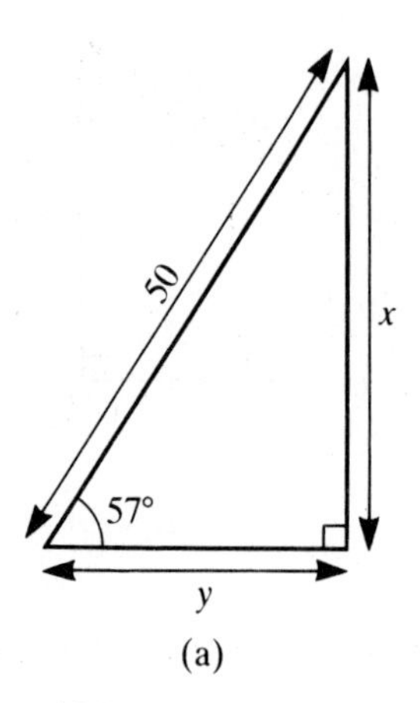

Figure 15.7 For example 1.

Solution

(a) $\sin 57° = \dfrac{\text{opposite}}{\text{hypotenuse}} = \dfrac{x}{50}$, i.e. $\sin 57° = \dfrac{x}{50}$

so in multiplying both sides of this equation by 50, we have

$$50 \times \sin 57° = \frac{x}{\cancel{50}_1} \times \cancel{50}^1$$

i.e. $x = 50 \sin 57° = 50 \times 0.8387 = 41.935$ *Ans*

$$\cos 57° = \frac{\text{adjacent}}{\text{hypotenuse}} = \frac{y}{50},$$

i.e. $\dfrac{y}{50} = \cos 57°$

so $y = 50 \times \cos 57° = 50 \times 0.5446 = 27.23$ *Ans*

(b) In the right-angled triangle XYZ,

$$\tan 42° = \frac{\text{opposite}}{\text{adjacent}} = \frac{x}{25}$$

so $x = 25 \times \tan 42° = 25 \times 0.9004 = 22.51$ *Ans*

In the right-angled triangle XYP,

$$\sin 50° = \frac{\text{opposite}}{\text{hypotenuse}} = \frac{x}{y}$$

and as $x = 22.51$,

$$\sin 50° = \frac{22.51}{y}$$

On multiplying both sides by y, we have

$$y \sin 50° = \frac{22.51}{\cancel{y}} \times \cancel{y}$$

and dividing by $\sin 50°$,

$$y = \frac{22.51}{\sin 50°} = \frac{22.51}{0.7660} = 29.39 \quad \textit{Ans}$$

2 The angle of elevation of the top of a building from a point at ground level and an unknown distance away is 24°. On moving 140 metres from this point directly towards the building the angle of elevation is remeasured at 70°. Determine the height of the building

Note: The *angle of elevation* is the angle between the horizontal and the line joining the observed point (in this case the top of the building) to the point at which measurement is made, see Figure 15.8(a).

The term *angle of depression* is the angle between the line of sight and the horizontal when an object is viewed from a point at a higher level, see Figure 15.8(b).

Solution

A diagram summarizing the data given for the problem is drawn in Figure 15.9. p = height of building and b = distance from building base to 70° elevation point, are marked in.

The problem is first to find equations linking these two unknowns.

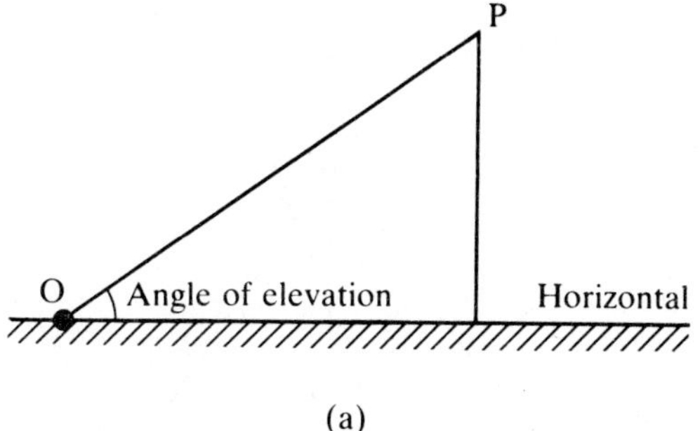

(a)

Angle of elevation of point P with respect to point O

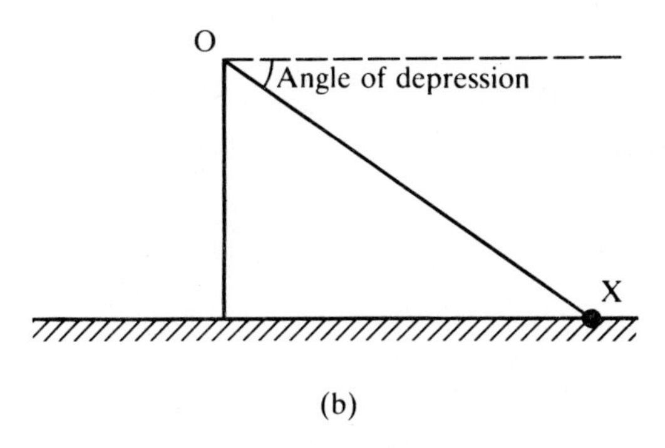

(b)

Angle of depression of point X with respect to point O

Figure 15.8.

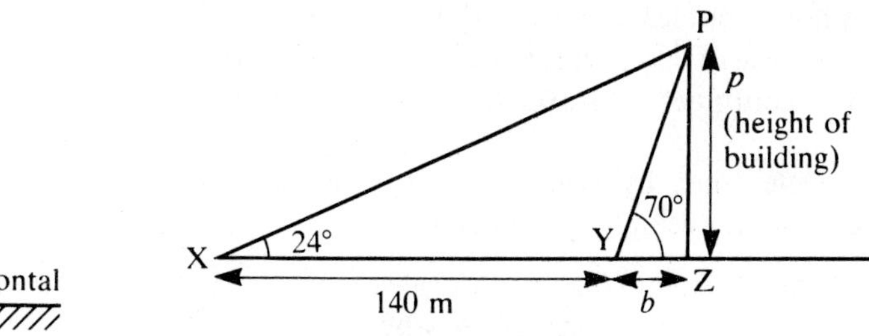

Figure 15.9 Diagram for example 2.

In the right-angled $\triangle$ PYZ.

$$\tan 70^\circ = \frac{p}{b}, \quad \text{so} \quad p = b \tan 70^\circ \qquad (1)$$

In the right-angled $\triangle$ PXZ,

$$\tan 24^\circ = \frac{p}{140 + b}, \quad \text{so } p = (140 + b) \tan 24^\circ \qquad (2)$$

From (1) $p = b \tan 70^\circ = 2.747b$,
as $\tan 70^\circ = 2.747$
so $b = p/2.747 = 0.364p$

Substituting for b in (2), we have

$$p = (140 + 0.364p) \tan 24^\circ$$
$$p = 0.445(140 + 0.364p),$$

as $\tan 24^\circ = 0.445$

$$p = 62.3 + 0.162\,p$$
$$p - 0.162p = 0.838p = 62.3$$
$$p = \frac{62.3}{0.838} = 74.3\,\text{m} \quad \textit{Ans}$$

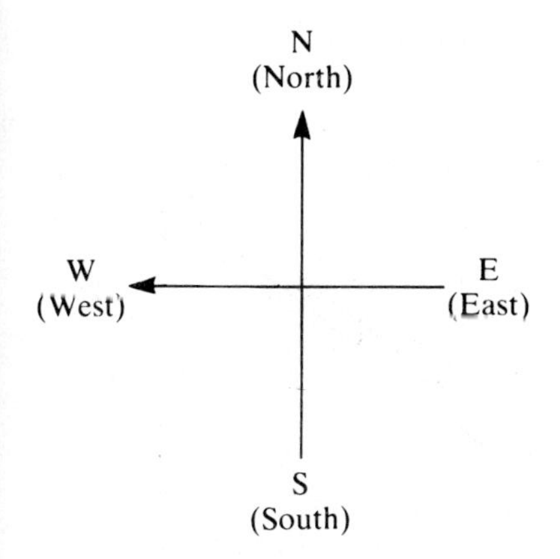

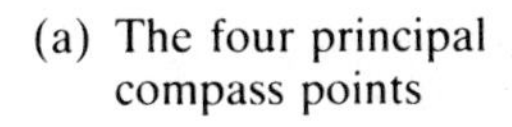

(a) The four principal compass points

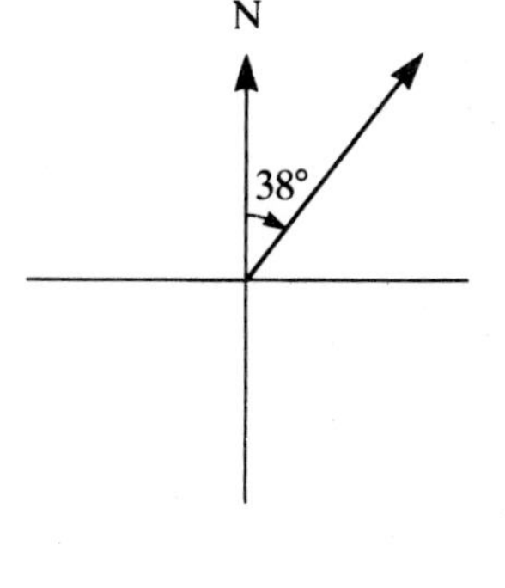

(b) Bearing 038° (38° clockwise from North)

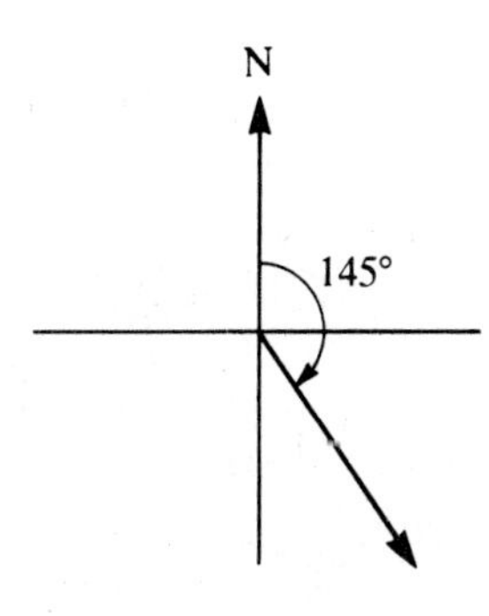

(c) Bearing 145°

Figure 15.10.

3 A hovercraft leaves port A and travels 50 km on a bearing of 038°. The craft then changes course and continues to travel on a new bearing of 145°.

Determine, by scale drawing and also by calculating using trigonometry:

(a) the maximum distance travelled northwards;
(b) the distance from port A when the craft is due east of A.

Note: In navigation, angular directions are defined with respect to the due north direction and are known as **bearings**. A bearing defines the direction by the angle measured from due north in the clockwise direction to the direction of travel or line of sight to a given point, see Figure 15.10.

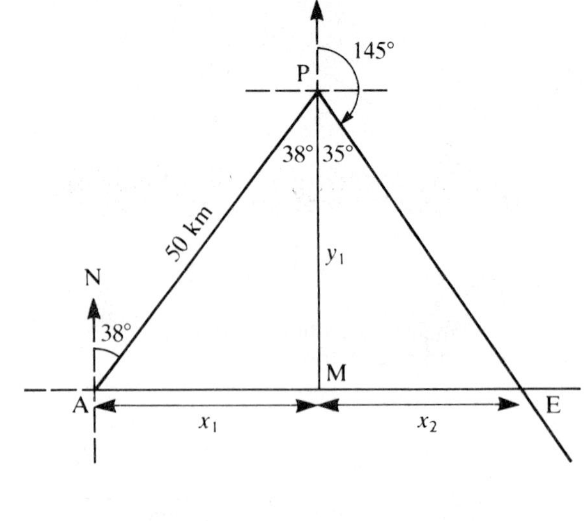

Scale 1 mm = 1 km

Figure 15.12 Diagram for Qu 0.

Solution

The scale drawing of the hovercraft's journey is drawn in Figure 15.11 using a scale 1 mm = 1 km. From this drawing:

(a) Maximum distance travelled north,
$MP = 39\,\text{km}$ *Ans*

(b) Distance from port A when due east,
$AE = 59\,\text{km}$ *Ans*

Using trigonometry:

(a) In right-angle triangle APM,

$$\cos 38^\circ = \frac{MP}{50}, \quad \text{so} \quad MP = 50\cos 38^\circ$$

$$\text{so} \quad MP = 50\cos 38^\circ = 50 \times 0.788 = 39.4\,\text{km} \quad \textit{Ans}$$

(b) In $\triangle$ APM

$$\sin 38^\circ = \frac{x_1}{50}$$

so $x_1 = 50\sin 38^\circ = 50 \times 0.6157 = 30.8\,\text{km}$

In $\triangle$ PME

$$\tan 35^\circ = \frac{x_2}{PM}$$

$$x_2 = PM \times \tan 35^\circ = 39.4 \times 0.7002 = 27.6\,\text{km}$$

Hence distance from port A when due east,

$$AE = x_1 + x_2 = 30.8 + 27.6 = 58.4\,\text{km} \quad \textit{Ans}$$

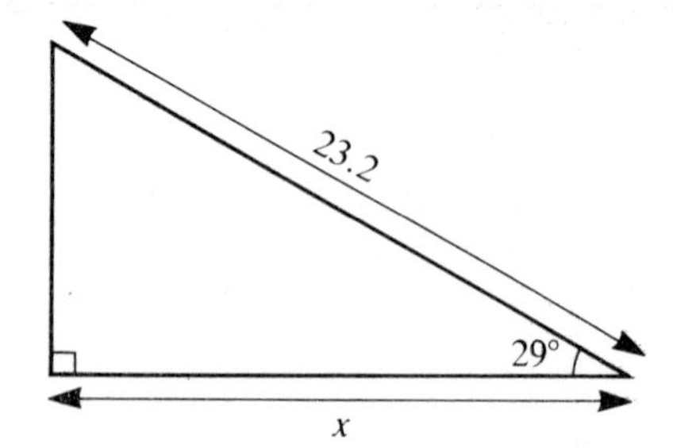

Figure 15.11 Scale drawing for example 3.

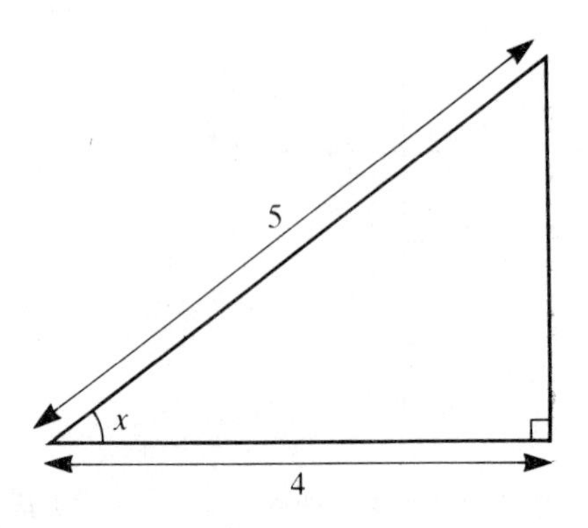

Figure 15.13 Triangle for Qu 2.

Test and problems 15

Multiple choice test: MT 15

Answer block:

Question No.	0	1	2	3	4	5	6	7	8
Answer	c								

Enter your answer, that is a, b, c or d in the column under the question number in the answer block above.
Note that Qu. 0 has already been worked out and the answer inserted.

Qu. 0 Determine the length x in the diagram of Figure 15.12.
Ans (a) 11.3 (b) 18.5 (c) 20.3 (d) 12.9

Solution

$$\cos 29^\circ = \frac{\text{adjacent}}{\text{hypotenuse}} = \frac{x}{23.2}$$

so $x = 23.2 \cos 29^\circ = 23.2 \times 0.8746 = 20.3$ *Ans*

Hence (c) is the correct answer and c is inserted under Qu. 0 in the answer block as shown.

Now carry on with the test.

Qu. 1 The value of $\cos 30^\circ$ is

Ans (a) $\frac{1}{2}$ (b) $\frac{\sqrt{3}}{2}$ (c) $\frac{1}{\sqrt{2}}$ (d) $\sqrt{3}$

Qu. 2 Determine the value of $\tan x$ in the triangle of Figure 15.13.

Ans (a) $\frac{4}{5}$ (b) 0.6 (c) 0.866 (d) $\frac{3}{4}$

Qu. 3 Determine the length x in the triangle of Figure 15.14.
Ans (a) 9.830 (b) 10.64
(c) 7.245 (d) 6.883

Qu. 4 Given $\sin\theta = \frac{5}{13}$, determine $\tan\theta$.

Ans (a) $\frac{13}{12}$ (b) 2.450 (c) $\frac{5}{12}$ (d) $\frac{12}{13}$

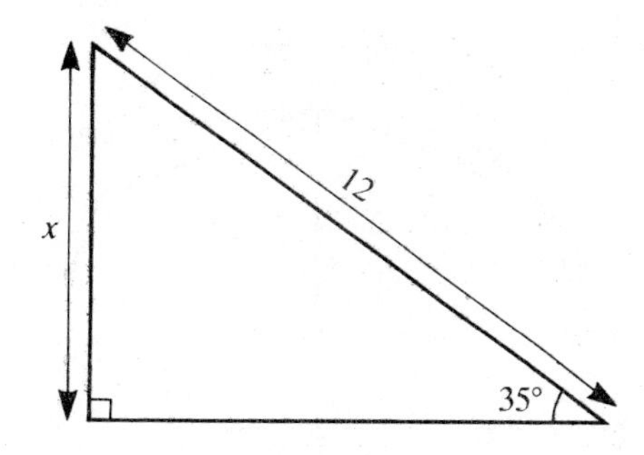

Figure 15.14 Triangle for Qu 3.

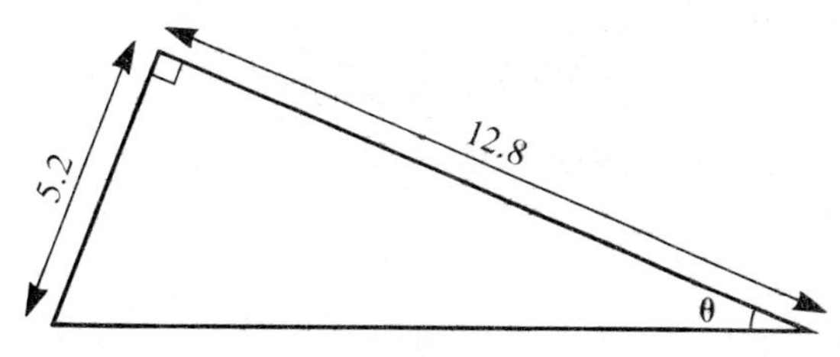

Figure 15.15 Diagram for Qu 5.

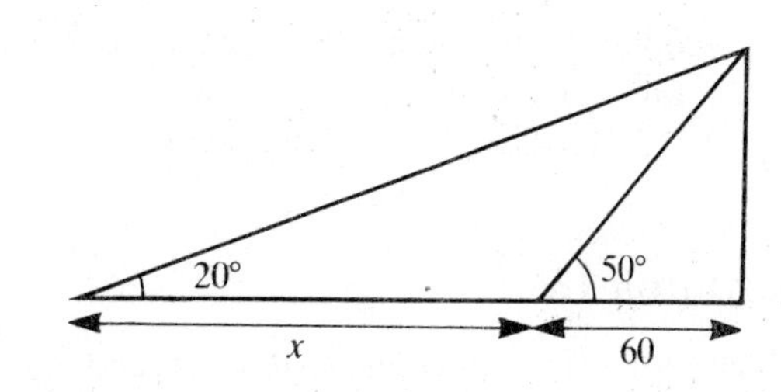

Figure 15.16 Diagram for Qu 7.

Qu. 5 Determine the angle θ in the triangle of Figure 15.15.
Ans (a) 24.0° (b) 23.0°
(c) 22.11° (d) 38.62°

Qu. 6 It is known that $\cos 53^\circ = 0.6018$. Use this result to determine $\sin 37^\circ$.
Ans (a) 0.7986 (b) 1.327
(c) 0.7536 (d) 0.6018

Qu. 7 Determine the length x in Figure 15.16.
Ans (a) 40 (b) 136.5
(c) 71.51 (d) 196.5

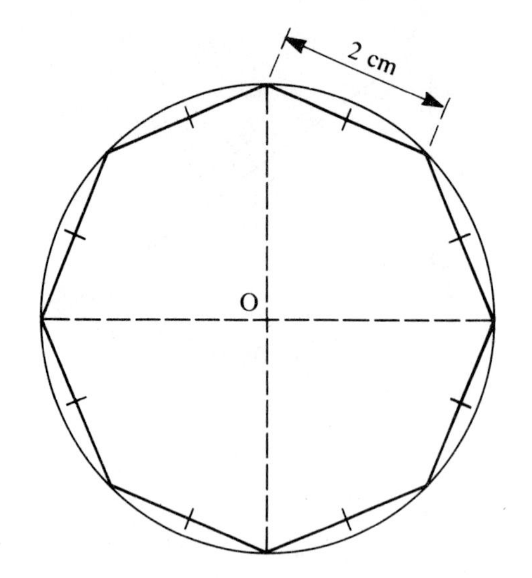

Figure 15.17 Diagram for Qu 8.

Qu. 8 Determine the radius of the circle in Figure 15.17. All the sides of the 8-sided figure contained by the circle are equal to 2 cm.
Ans (a) 2.613 cm (b) 2 cm
(c) 4 cm (d) 3.2 cm

Problems 15

1 Determine the values of $\sin\theta$, $\cos\theta$ and $\tan\theta$ for the angle θ in the triangles of Figure 15.18.

2 Use a calculator or 4-figure tables to determine:
(a) $\cos 42.3°$ (b) $\sin 53.4°$ (c) $\tan 79°$
(d) $\sin 14° 54'$ (e) $\tan 46° 2'$ (f) $\cos 32° 17'$

3 Using a calculator or tables determine:
(a) $\tan^{-1} 3.645$ (the angle whose tan is 3.645)
(b) $\sin^{-1} 0.9125$ (the angle whose sin is 0.9125)
(c) $\cos^{-1} 0.5257$ (the angle whose cos is 0.5257)

4 Determine without using tables or a calculator:
(a) $\cos 45°$, $\tan 45°$
(b) $\sin 30°$, $\cos 30°$, $\tan 30°$
(c) $\tan 60°$

5 Using the trigonometric identities:

$$\sin(90 - \theta) = \cos\theta, \quad \tan\theta = \sin\theta/\cos\theta$$

and $\sin^2\theta + \cos^2\theta = 1$, determine

(a) $\cos 65°$ given $\sin 25° = 0.4226$
(b) $\tan 32°$ given $\sin 32° = 0.5299$
(c) $\cos 78°$ given $\sin 78° = 0.9781$

6 Determine the unknown lengths marked-in on the diagrams of Figure 15.19.

7 The longer sides of a rectangular field are 70 m and a diagonal drawn across the field makes an angle of 41.5° with them. Calculate the length of the shorter sides and the area of the field.

8 The angle of elevation of the top of a building from a point 50 m from its base on level ground is 30°. Calculate the building height.

Calculate also the height of a second building whose base is 100 m from the first, given that the angle of depression from the top of the first to the top of the second is 5°.

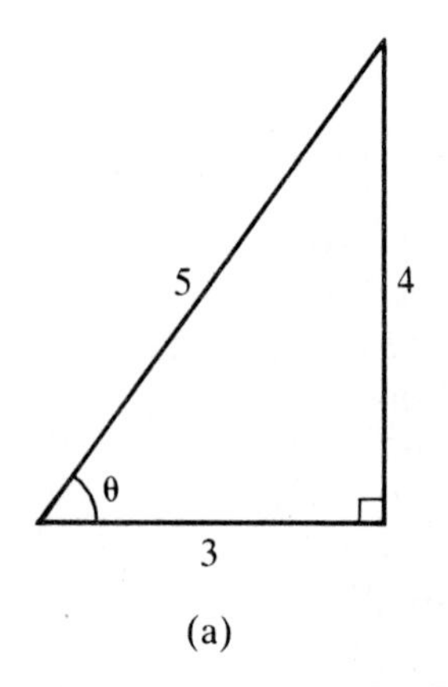

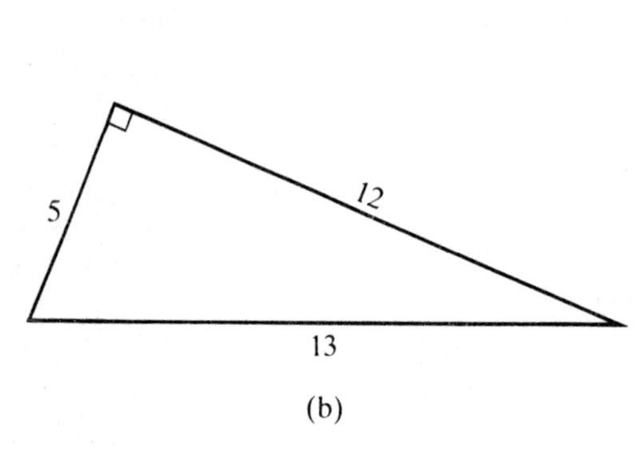

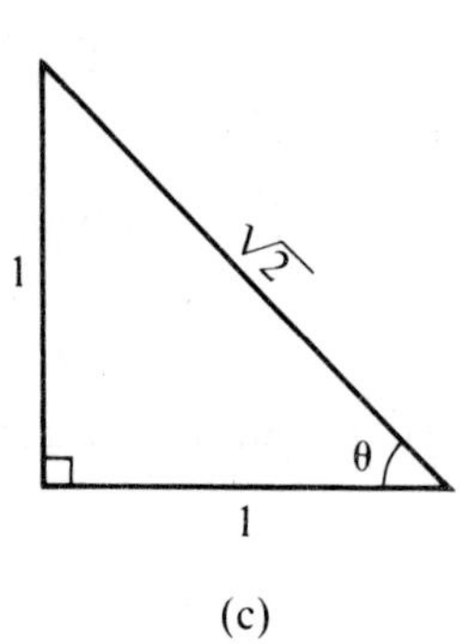

Figure 15.18 Triangles for problem 1.

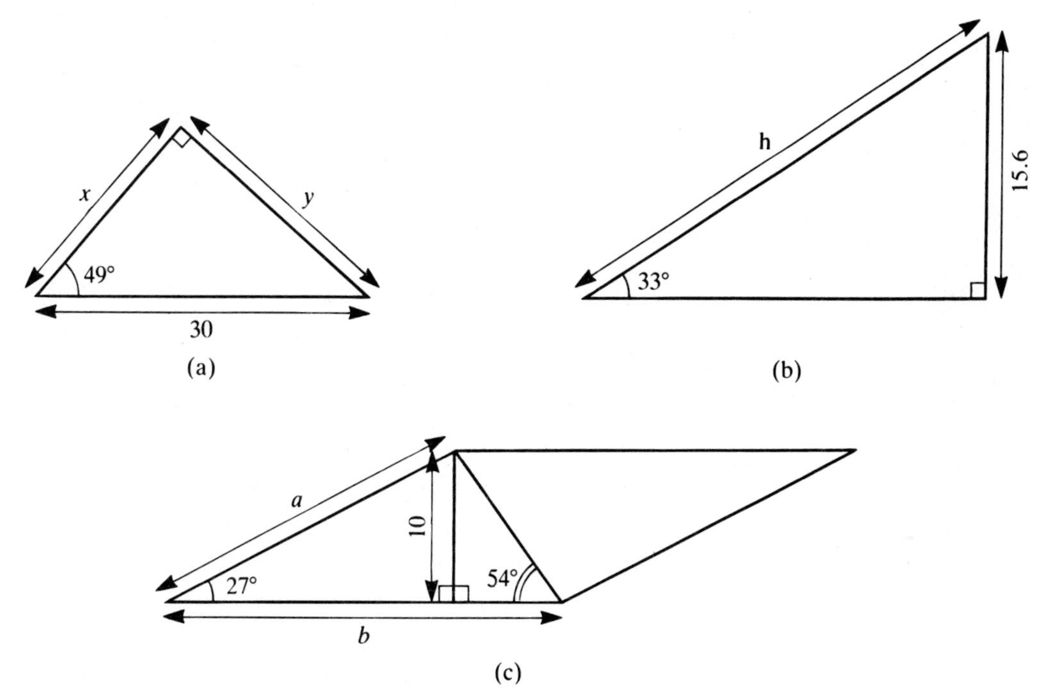

Figure 15.19 Diagrams for problem 6.

9 A yacht sails from a point X on a bearing of 055° for a distance of 10 km. Determine by scale drawing the distance travelled north and the distance travelled east.

A change of course is made and the yacht steers on a bearing of 130°. Determine by scale drawing the distance from X when the yacht has sailed 20 km on this course.

Check the accuracy of your solutions using trigonometry.

Part Four: Statistics

16 Introduction to statistics

General learning objectives: to collect, tabulate and summarize statistical data and interpret data descriptively.

16.1 Introduction to statistics

The term '*statistics*' is very familiar and very frequently used in our society. It is often used to denote sets of numerical data, normally presented in tabular or diagrammatic form, which record all forms of information – facts and figures relating to virtually all aspects of industrial, economic, scientific, medical, sociological, etc. matters.

More formally statistics can be defined as the study which deals with the collection, preparation, presentation, interpretation and analysis of quantitative data to aid the deduction of representative and meaningful conclusions, to indicate current trends and to attempt to predict future probabilities.

The study of statistics can be divided broadly into *descriptive statistics* and *mathematical statistics*. Descriptive statistics deals basically with the compilation and presentation of data to provide concise information on which to base decisions, while mathematical statistics is concerned primarily with the applications of probability theory.

In this chapter we are concerned only with descriptive statistics. We consider some basic ways by which data may be collected and processed. We show how statistics may be represented in tabular and diagrammatic form and how these diagrams may be interpreted to bring out the most significant features of the data.

16.2 Data collection: discrete and continuous data

The collection of data to compile statistics is carried out in numerous ways. On the one hand, collection may be a straightforward counting process whilst on the other, the collection may entail the computer logging of several sets of measurement data obtained in testing a complex system or even an entire industrial complex.

The raw data to be collected may be classed as primary or secondary. Primary data refer to data collected at source by making and recording measurements, by counting, by undertaking market surveys, etc. Secondary data refer to already existing data in some edited or published form obtained from government departments, research groups, etc.

Statistical data may be termed either continuous or discrete.

Continuous data can essentially take all values within a certain range and can normally be measured very accurately. The data values are 'continuous' and plotting the results could lead to a smooth-line graph. For example, if the temperature of a room is monitored and recorded every few minutes the resulting data would be more or less continuous and a graph of temperature versus time could be plotted and a smooth curve obtained.

Discrete data on the other hand cannot take all values and are normally measured in whole number units. Counting processes produce discrete data. In fact most statistical data are discrete, although sometimes the distinction becomes blurred.

The process of systematic counting is one of the simplest but most important means of collecting certain forms of statistical data. For example, in a traffic survey to assess the number of vehicles using a given route, the number of cars, lorries, motorcycles, etc. may be counted and logged on a time sheet. This information, appropriately classified and presented, may then be used to aid the case for the construction of a bypass road.

A census to determine the population of a town or country is essentially a counting process. On a production line, completed articles may be counted electronically and the results stored and processed by a computer to record batch outputs; faults, delays, alarms, etc. may also be recorded for future analysis – all these processes are essentially counting processes producing discrete data.

16.3 Sampling and population

Very frequently it is neither feasible nor practical to collect and test every scrap of potentially valuable data. Clearly it could be both uneconomic and far too time-consuming to make tests on, for example, every component at each stage of its manufacture and log the results. Likewise, in making election surveys it would clearly be impossible to ask every voter for his/her views. In weather forecasting there is only a finite number of ground stations and a small number of satellites to collect data and yet we still expect – even if the forecasters stray sometimes – accurate predictions.

Many of the problems posed above may be overcome in a practical way be collecting data from representatives **samples** rather than all possible sources. **Sampling** is an essential part of market planning and research. In theory it allows us to learn about a 'whole' from the study of only a 'part'. Sampling can provide results which can be comparable with those obtained from an exhaustive study but provide the benifits of considerable savings in cost and time.

Obviously, for the information obtained to be of real value, the sample must be representative of the population. In the statistical sense, **population** is used to include any collection of things or people. A random sample is one in which each member of the total population has an equal chance of being selected. Sampling must be representative of the population yet free from bias.

Considerable thought must always be given to the optimum amount and quality of data that is required to be collected to produce meaningful and reliable statistics. Insufficient data could at best lead to rather superficial conclusions. Over-collection of data obscures the real objectives and leads to extra work in extracting meaningful statistics. So often nowadays, statistics are collected with little or no reason and the question should always be asked, 'Why am I collecting these statistics and what do I really require them for?'

The collection of data using samples is an extremely powerful method, but it must be stressed that great skill is required in assessing what constitutes the correct size of a sample so that it is both representative and free from any bias. For example, in each batch of, say, 20 000 components from a production line a sample of 100 may be selected at random and then tested. The results of this sample alone are then used to estimate the quality of the complete batch and to 'correct' production if any faults are found in the test sample.

In election forecasting, representative samples of the total population of voters are canvassed for their views; for example, a survey may be based on selecting 5000 people drawn from 40 or so distinctive areas of the country. If these samples were truly representative and the voters did not change their minds on election day, then accurate election predictions could be made.

Great care must be exercised in certain areas. For example, the testing of a new drug may require several years of study with very careful analysis of results and the widest possible sampling to cover all conceivable cases. Clearly it could be negligent to test a drug which may reach several millions of people without exhaustive tests.

16.4 Preparing data for analysis: grouping, frequency distribution, tally diagrams

After the collection of data, the facts and figures should be presented in a way that enables them to be easily interpreted.

One extremely useful and very commonly used means is to group the data into a number of **class intervals** – sub-divisions of the total **range** between maximum and minimum data values – and then to tabulate the information in a **frequency distribution** table or display the sorted data graphically. Class intervals are usually chosen to be of equal size, except for lower and upper-end extremes, and actual size will normally be determined by the overall data range and the number of data values involved.

The **tally diagram** provides an easy way of counting the number of times – the **frequency** – that data values fall within a given class interval. The use of a tally diagram and the formation and presentation of results in a frequency table is illustrated in the following example.

Example

A sample of 40 cans is selected from a production batch of nominal weight 100 grams. The weight of each can is measured and the results are rounded down to the nearest gram, e.g. 102.9 g is recorded as 102 g, and are listed below.

105	97	100	95	104	93	105	98
100	99	100	95	99	97	104	102
102	96	93	106	100	99	100	94
105	102	97	101	94	101	91	105
100	100	96	104	97	104	97	101

Use a tally diagram to group the can weights into class intervals of 5 grams. Using the results of the tally diagram form a frequency table summarizing the frequency distribution information of can weights, i.e. a table showing the class intervals and number of cans in each interval.

Procedure

1 First determine the *range*. The highest and lowest weights are ringed in the data list and are 106 and 91.

2 Select appropriate class interval ranges to cover the total range 91 to 106. In this problem we are told to select a 5 gram interval so the following class intervals are appropriate:

90–94, 95–99, 100–104, 105–109

Note: the class interval 90–94 includes cans with weights equal to or greater than 90 but less than 95, i.e. the true class boundary limits are 90.0 to 94.99. Likewise the class interval 95–99 includes all can weights between 95.0 to 99.99... and so on. Each class interval covers an effective range of 5 g with all recorded weights rounded down to the nearest gram.

3 Now count and record the number of can weights in each class interval. This information may be extracted from the unsorted list by forming a **tally diagram** as shown below:

Weight class interval	*Tally marks*	*Number of cans*
90–94	卌	5
95–99	卌 卌 ///	13
100–104	卌 卌 卌 //	17
105–109	卌	5

Work systematically through the list of data row by row and make an oblique stroke – the 'tally mark' – for each value against the relevant class interval in the tally diagram. After four marks the fifth is normally denoted by a horizontal line to indicate a sub-total of 5 and facilitate final totalling.

4 After totalling the tally marks for each class interval the frequency table may be constructed, i.e.

Frequency distribution table

Class interval (grams)	90–94	95–99	100–104	105–109
Frequency (no. of cans)	5	13	17	5

The frequency distribution table, although losing some finer points of detail, presents the data in a far more informative way than the initial un-ordered list. We can see at a glance that $13 + 17 = 30$ cans have weights within $\pm 5\%$ of the nominal 100 g value and that all the cans in the sample of 40 lie within $\pm 10\%$.

Further it would not be unreasonable to make the following estimates since the 40 cans were selected at random:

1 Most of the batch would have weights within $\pm 10\%$ of 100 g.
2 $30/40 = 3/4$ or 75% of the cans should be within $\pm 5\%$.

Thus if the batch contained 10 000 cans we could make the tentative prediction that the majority of cans would be within $\pm 10\%$ and $3/4 \times 10\,000 = 7500$ would be within $\pm 5\%$. It must be stressed that these 'predictions' are very approximate since our sample of 40 out of 10 000 is small and although selected at random may not be totally representative.

16.5 Frequency and relative frequency

The term **frequency** as used in statistics is defined with respect to a class interval as the number of items (results, values, etc.) which fall within the limits defining the class interval.

The term **relative frequency** is the frequency of a class interval expressed as a fraction of the total number of items in the complete data list. **Relative percentage frequency** is the frequency expressed as a percentage of the total number of items, i.e.

$$\text{relative frequency} = \frac{\text{frequency of given class}}{\text{total no. of items}}$$

relative percentage frequency

$$= \frac{\text{frequency of given class}}{\text{total no. of items}} \times 100$$

Thus, for our previous example of 40 cans:

Class interval	Frequency	Relative frequency	Relative % frequency
90–94	5	$5/40 = 0.125$	12.5%
95–99	13	$13/40 = 0.325$	32.5%
100–104	17	$17/40 = 0.425$	42.5%
105–109	5	$5/40 = 0.125$	12.5%

Frequency is, of course, dependent on the size selected for the class interval; frequency (no. of items) will obviously increase as the class interval increases. It is possible to have a zero frequency (or many zeros) where no values occur in a given class interval. It is up to us to select suitable class intervals to aid the presentation of frequency distribution data and help interpretation. Too small a class interval may lead to finer detail than is required and may actually obscure general trends. Too coarse a class interval may cause important characteristics to be overlooked.

Example

In a reliability test on a sample of 50 motors taken from a production batch of 1000 the following life-times (running time in hours before breakdown) were recorded:

930	900	912	886	905
884	938	885	890	897
908	865	908	877	927
888	898	891	900	910
920	873	914	868	935
905	892	914	911	882
907	891	901	913	917
927	862	882	916	898
922	895	890	875	945
891	876	879	901	871

Note: life-times rounded down to nearest hour, e.g. 907.8 is recorded as 907.

Using a class interval size of 10 hours, present the above results in a frequency table and calculate the relative percentage frequency of the most and least densely populated class.

Solution

The minimum and maximum life-times in the data list are respectively 862 and 945 so suitable 10-hour class intervals would be 860–869, 870–879.... 940–949. Using a tally diagram to sort and count the data into these classes, we obtain:

Class interval (life-time, hours)	Tally marks	Frequency
860–869	///	3
870–879	~~////~~ /	6
880–889	~~////~~ /	6
890–899	~~////~~ ~~////~~	10
900–909	~~////~~ ////	9
910–919	~~////~~ ///	8
920–929	////	4
930–939	///	3
940–949	/	1
	Total	50

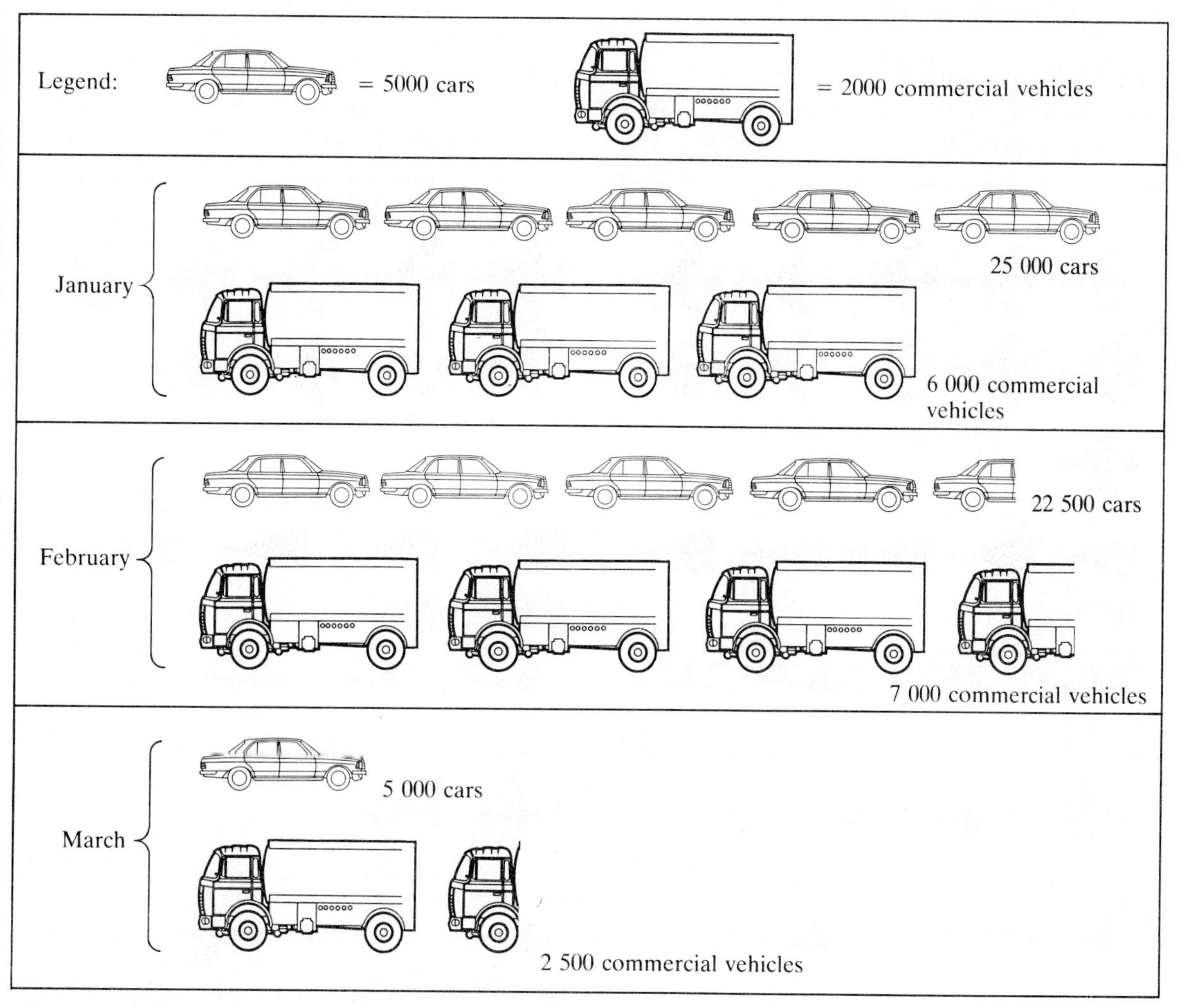

Figure 16.1 Pictogram showing monthly production figures for a three-month period.

The most densely populated class interval is 890–899 hours with frequency = 10 and hence

relative percentage frequency

$$= \frac{\text{frequency}}{\text{total no. items}} \times 100$$

$$= \frac{10}{50} \times 100 = 20\%$$

The least-populated class interval is 940–949 with only 1 item, so relative percentage frequency $= \frac{1}{50} \times 100 = 2\%$.

16.6 Pictorial and diagrammatic presentation of statistical data

16.6.1 Pictograms

One of the most visual and easily interpreted means of displaying statistical data and the one extensively used by newspapers and television is the pictogram. In a **pictogram** a picture or symbol is used to depict a numerical value or statistical figure/fact.

Figure 16.1 shows a pictogram illustrating the production of a motor vehicle manufacturer over a three-month period. A picture of a car is used to denote a production unit of 5000 cars and fractional pictures to denote fractional units, e.g. half a car 'picture' denotes 2500 cars; by contrast commercial vehicle production is depicted by a transporter picture, each full picture denoting 2000 vehicles.

Figure 16.2 illustrates the use of pictograms to both represent and compare data – in this case the strength of two airforces.

16.6.2 100% bar charts and pie diagrams

Two useful and illustrative ways of presenting frequency distribution data in diagrammatic form are 100% bar charts and pie charts or diagrams.

The **100% bar chart** consists of a column subdivided into rectangular blocks, where the length or area of each block represents the relative fre-

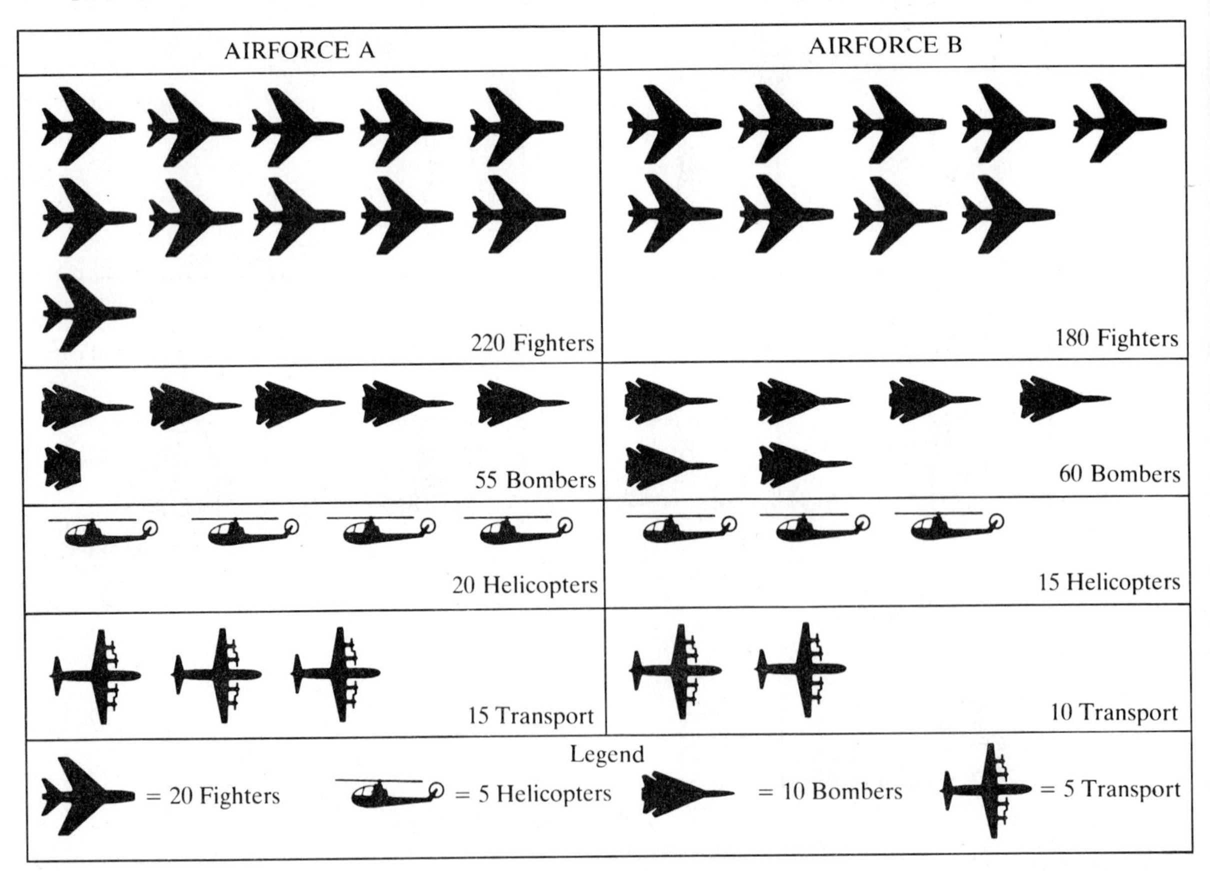

Figure 16.2 Pictogram showing strengths of two air forces.

quency (fractional number of items) of a particular class interval. The blocks may be shaded or coloured to increase the visual effect. Figure 16.3 shows the 100% bar chart of the machine life-time example of section 16.5 whose frequency distribution is summarized below:

Class interval (life-time range in hours)	Frequency (no. of items per class)	Relative percentage frequency (%)
860–869	3	6
870–879	6	12
880–889	6	12
890–899	10	20
900–909	9	18
910–919	8	16
920–929	4	8
930–939	3	6
940–949	1	2

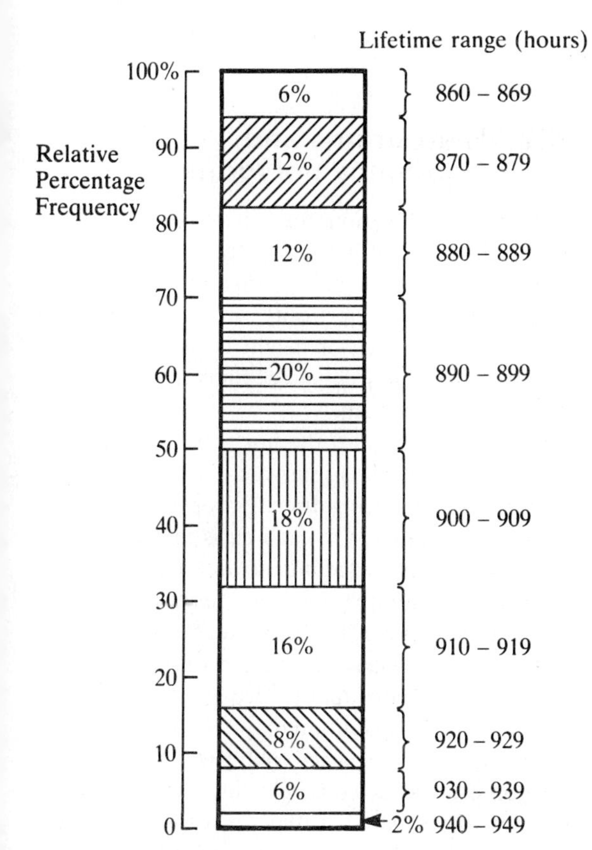

Figure 16.3 100% bar chart presenting results of life-time tests on 50 motors.

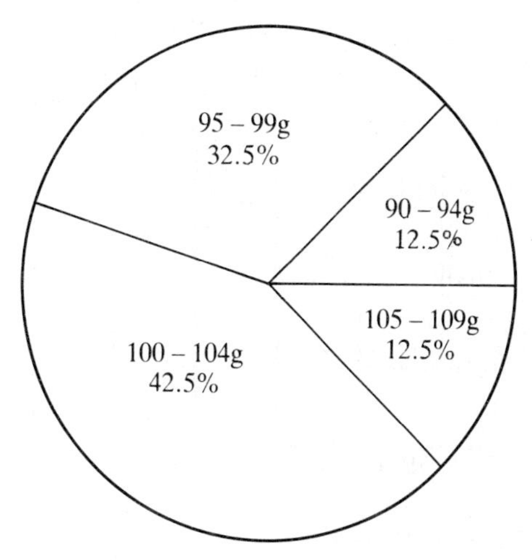

Figure 16.4 Pie chart showing distribution of weights in a sample of 40 cans.

In the **pie chart** or **pie diagram** representation, a circle (the pie) is sub-divided into sectors, one sector for each class interval. The area of the individual sectors, and therefore the angle subtended at the circle centre, is made directly proportional to the relative frequency of the class interval. For example, the pie diagram showing the weight distribution of our 40 can example (see sections 16.4 and 16.5) and based on the following data is drawn in Figure 16.4.

Class interval (grams)	Frequency	Relative frequency	Pie sector angle (ref. frequency × 360°)
90–94	5	5/40 = 0.125	0.125 × 360 = 45°
95–99	13	13/40 = 0.325	0.325 × 360 = 117°
100–104	17	17/40 = 0.425	0.425 × 360 = 153°
105–109	5	5/40 = 0.125	0.125 × 360 = 45°

The pie diagram is constructed using a protractor to mark out sector angles which are calculated from the relative frequency values as shown in the above table.

16.6.3 Vertical and horizontal bar charts

Vertical and horizontal bar charts are also used extensively to present 'grouped' data. The length of each bar corresponds to the number of items or value of its corresponding class, the only difference between the two forms of display is that the bars

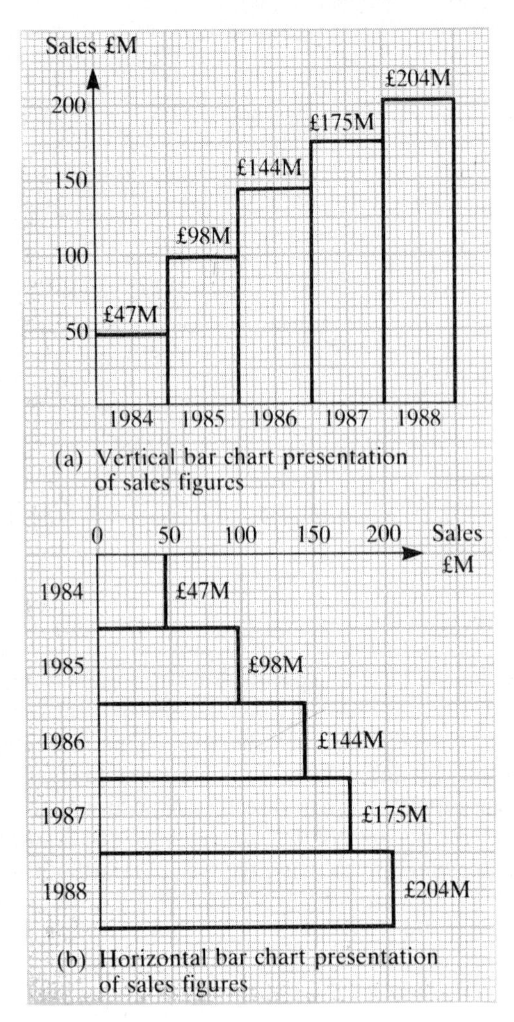

Figure 16.5.

are drawn vertically in a vertical bar chart and horizontally in a horizontal bar chart.

These charts are especially useful when trends are to be emphasized. For example, the vertical and horizontal bar chart presentations for five consecutive years of sales of a company:

Year	Total sales in £ million
1984	47
1985	98
1986	144
1987	175
1988	204

are shown in Figure 16.5. The length of each bar is made directly proportional to the respective year's sales figure. Both charts show clearly that the sales of the company are steadily increasing and so from a sales point of view the company appears to be doing very well.

16.7 Histograms and frequency polygons

Frequency distributions can also be represented graphically by plotting along the horizontal axis the intervals into which the range of data values is split, and drawing vertically on each interval a rectangle whose area or height is proportional to the class interval frequency. Such a diagram is known as a **histogram**.

Thus in a histogram the horizontal axis is used as a base for the class intervals, the width of each interval normally being directly proportional to the class interval range. On each class interval base a rectangle is constructed whose area represents the frequency (the number of values, items, etc.) of the class interval. When the class intervals are equal, which is often the case, the heights of the rectangles are directly proportional to frequency.

A frequency distribution may also be represented by drawing a **frequency polygon**. A frequency polygon is constructed by plotting frequency vertically at the centre of the corresponding class intervals. Thus the polygon is similar to the histogram, but instead, frequency is plotted as a series

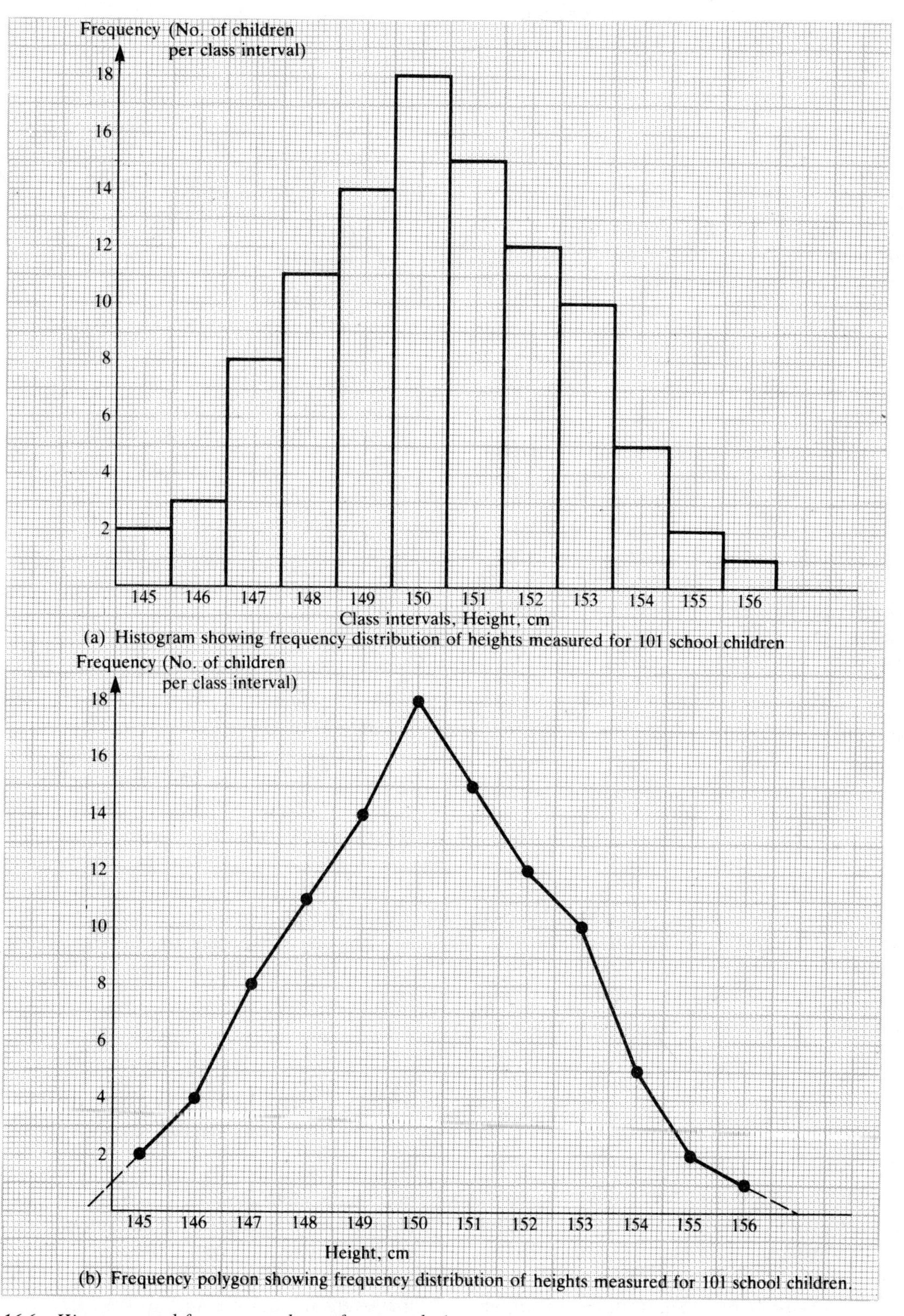

Figure 16.6 Histogram and frequency polygon for example 1.

of points at each class interval centre. The polygon is completed by joining consecutive frequency points with straight lines.

Examples

1 The following frequency table summarizes the heights measured to the nearest centimetre for 101 school children. Plot a histogram and a frequency polygon to show these results graphically.

Class interval (Height range, cm)	Interval centre (Height, cm)	Frequency (No. of children)
144.5–145.5	145	2
145.5–146.5	146	3
146.5–147.5	147	8
147.5–148.5	148	11
148.5–149.5	149	14
149.5–150.5	150	18
150.5–151.5	151	15
151.5–152.5	152	12
152.5–153.5	153	10
153.5–154.5	154	5
154.5–155.5	155	2
155.5–156.5	156	1

Note: heights are measured to the nearest centimetre so, for example, a height of 144.5 cm would be rounded up to 145 cm while a height of 144.49 would be rounded down to 144 cm. The class intervals are 1 cm wide and 145 cm, 146 cm ... 156 cm represent the class interval centre.

Solution

The histogram and frequency polygon presenting the above frequency table information are constructed and shown in Figure 16.6(a) and (b), respectively.

2 Two companies produce identical products. Samples of 1000 are taken from each company and tested independently. The following frequency table summarizes the life-times.

Class interval (Life-time hours)	Frequency (no. per class) Company A	Company B
below 900	26	153
900–1000	114	426
1000–1100	350	84
1100–1200	375	62
1200–1300	62	28
1300–1400	41	130
above 1400	32	117
Total	1000	1000

Draw the histogram to present this information and compare the product supplied by company A and B.

Solution

The histograms of frequency versus life-time for the two companies A and B are plotted in Figure 16.7(a) and (b), respectively. Some 'poetic licence' has been taken in representing the 'below 900 hours' and 'above 1400 hours' classes. Both have been represented with the same class interval base length.

On making a comparison of the two histograms it is seen that the life-time of B's product is more erratic, with a greater number of failures below 900 and much greater failures in the 900–1000 hour class interval, but with marginally better performance than A's above 1300 hours. For more consistent performance company A appears to offer better reliability than company B.

16.8 Cumulative frequency and the ogive curve

Histograms, frequency polygons and other diagrammatic charts are all extremely useful to show at a glance the salient points relating to the frequency distributions of graphs of data.

An additional method which is especially useful for ascertaining a measure of the spread of values over the complete data range is to plot a cumulative frequency curve. The **cumulative frequency** is the

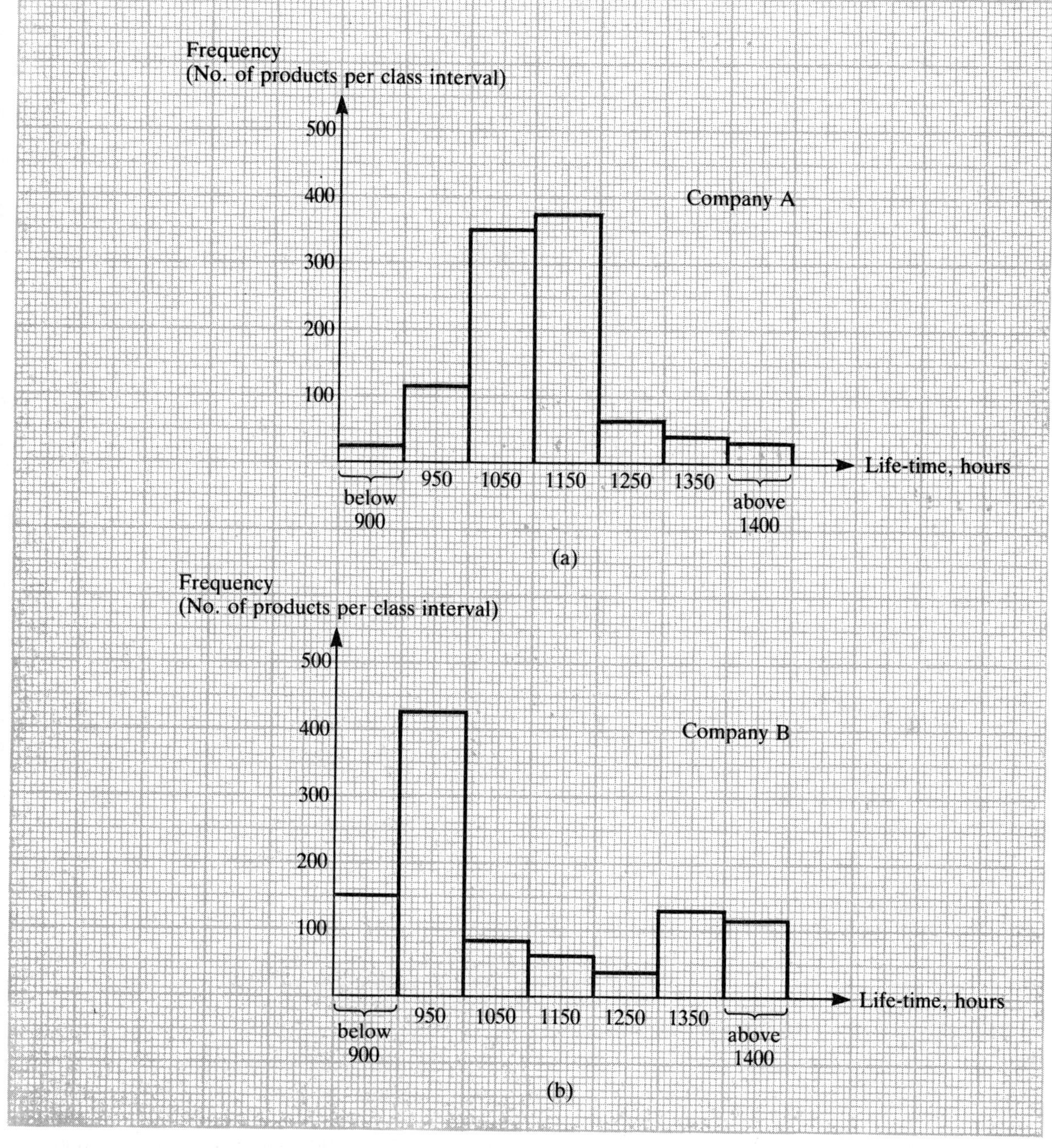

Figure 16.7 Histograms for example 2.

sum of the frequencies of the class intervals beginning at the first up to and including the present interval and the **cumulative frequency distribution curve** is obtained by plotting cumulative frequency (y-axis) versus class range (horizontally or x-axis).

For example, the cumulative frequency table for our previous example for school-children's heights is given below:

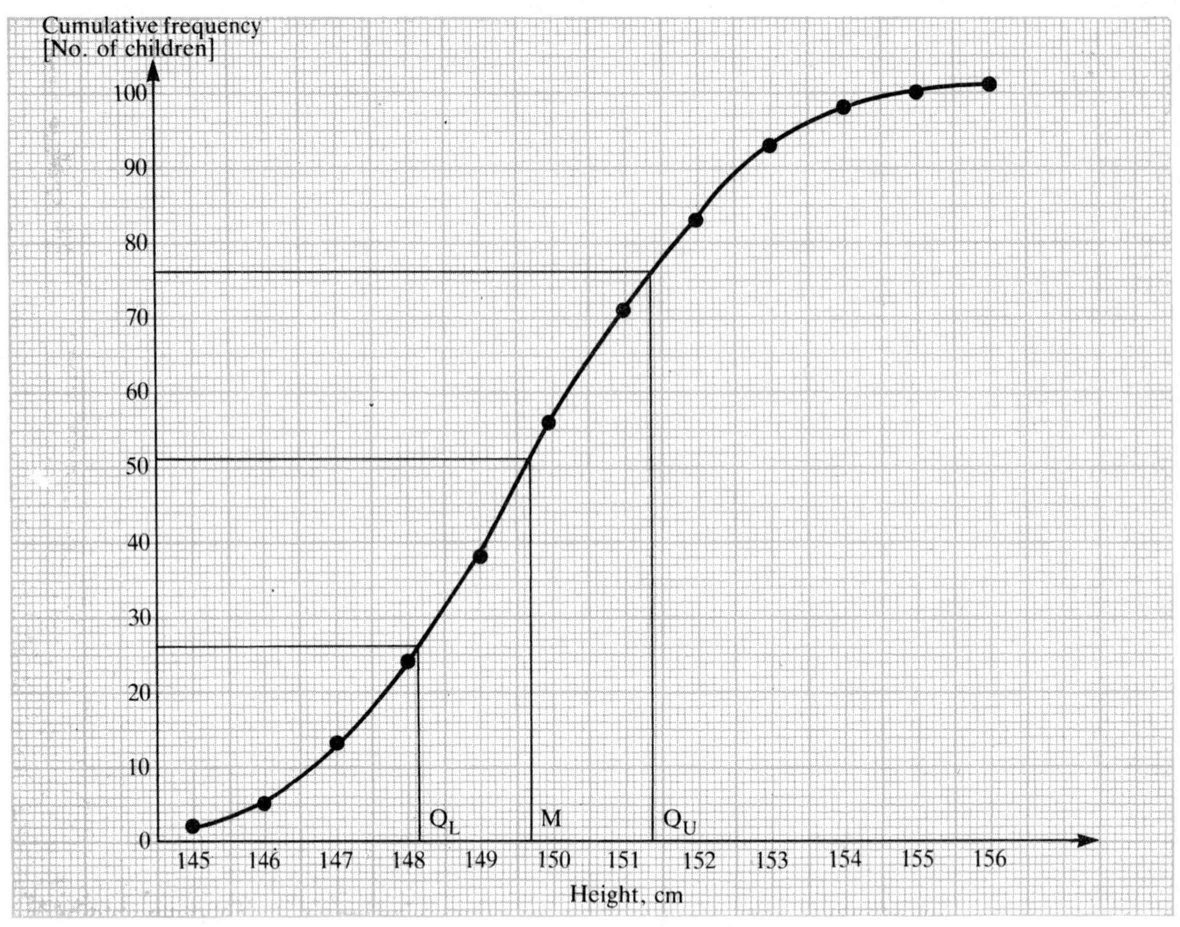

Figure 16.8 Cumulative frequency curve for 101 school children showing distribution of heights:
M is the median (middle) value, $M = 149.7$ cm
Q_U is upper quartile value, $Q_U = 151.3$ cm
Q_L is the lower quartile value, $Q_L = 148.2$ cm
semi-interquartile range $= \frac{1}{2}(Q_U - Q_L) = 1.55$ cm.

Class interval centre (height, cm)	Frequency	Cumulative frequency
145	2	2
146	3	5
147	8	13
148	11	24
149	14	38
150	18	56
151	15	71
152	12	83
153	10	93
154	5	98
155	2	100
156	1	101

The corresponding cumulative frequency curve is plotted in Figure 16.8.

A cumulative frequency curve, because of its characteristic shape is sometimes called the **ogive** curve. Ogive is the French term for a Gothic arch.

Cumulative frequency curves may be used to determine the following quantities which are important in statistical analysis:

1 The **median**

The median is the middle value of the data list when the list is arranged in order of magnitude. The median divides the frequency distribution into two equal halves.

In Figure 16.8 the median, marked *M*, corresponds to the 51st term, the middle of the 101 terms.

Reading directly from the graph: median $M = 149.7$ cm.

2 The **upper and lower quartiles**
The quartile values together with the median divide the frequency distribution into four equal parts. The lower quartile is the value one-quarter through the distribution, the upper quartile is the value three-quarters through the distribution. In Figure 16.8 the lower quartile Q_L corresponds to the 26th term and the upper quartile Q_U to the 76th term.
Reading off from the curve, we obtain: $Q_L = 148.2$ cm; $Q_U = 151.3$ cm.

3 The **interquartile range**
Fifty per cent of the values in a cumulative frequency distribution lie between the upper and lower quartiles. The range $Q_U - Q_L$ is known as the interquartile range and $\frac{1}{2}(Q_U - Q_L)$ as the semi-interquartile range. The latter is very often used as an estimate of the spread or dispersion of the data values.
The interquartile range for Figure 16.8 is:

$$Q_U - Q_L = 151.3 - 148.2 = 3.1 \text{ cm}$$

and the semi-interquartile range:

$$\frac{1}{2}(Q_U - Q_L) = 1.55 \text{ cm}$$

so 50% of the children have heights ± 1.55 cm about the median value of 149.7 cm.

Test and problems 16

Multiple choice test MT: 16

Answer block:

Question No.	0	1	2	3	4	5	6	7	8
Answer	c								

Enter your answer, that is a, b, c or d in the column under the question number in the answer block above. Note that question Qu. 0 has already been worked out and the answer has been inserted.

Qu. 0 The life-time of electrical components is summarized in this table:

Life-time (hours)	Frequency
2000–2099	8
2100–2199	14
2200–2299	23
2300–2399	27
2400–2499	16
2500–2599	12

Determine the relative frequency of the most populated life-time class.
Ans (a) 54% (b) 0.23% (c) 27% (d) 8%

Solution
The most populated class or group is 2300–2399 hours with a frequency of 27, i.e. 27 components have a life-time between 2300 and 2399 hours and this exceeds all other classes.
The total for all classes is $8 + 14 + 23 + 27 + 16 + 12 = 100$, so the relative frequency for the 2300–2399 hour class is

$$\frac{27}{100} = 0.27 \quad \text{or} \quad 27\%$$

Hence (c) is the correct answer and 'c' is inserted under Qu. 0 in the answer block as shown

---

Now carry on with the test:

Qu. 1 Figure 16.9 shows the pie chart for the employees of a certain company employing a total of 720 women and men. How many staff are employed in administration?
Ans (a) 100 (b) 200 (c) 180 (d) 300

Qu. 2 The lengths of 200 rods measured and rounded down to the nearest centimetre are listed in the following frequency table:

Length (cm)	46	47	48	49	50	51	52	53	54
Frequency	12	14	15	24	40	22	27	20	26

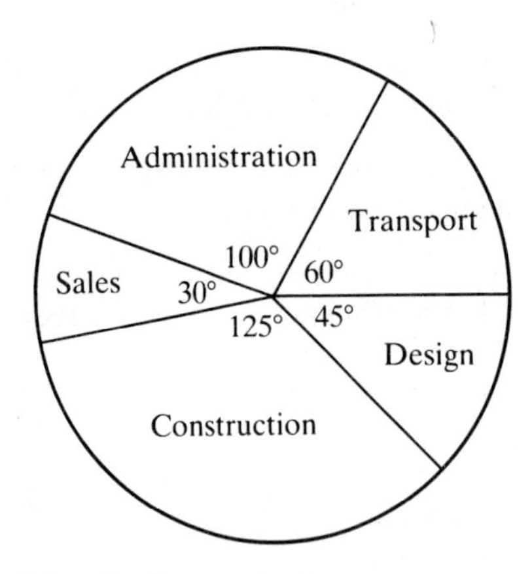

Figure 16.9 Pie diagram for Qu 1.

What percentage of rods are 50 cm or greater in length?
Ans (a) 33% (b) 57% (c) 50% (d) 67.5%

Qu. 3 The following data refers to the life-times of a sample of 50 light bulbs taken from a production batch:

Life-times in 100 hour units (rounded to nearest 100 hours)

9	15	2	10	9	7	12	7	11	1
12	10	14	12	22	8	11	7	14	11
14	11	4	20	11	10	6	12	10	10
9	15	10	4	14	10	8	11	13	2
12	9	18	12	11	11	10	22	7	15

Which of the following frequency tables accurately represents the data?

Ans

CLASS	below 900 h	*Centre of class interval* 900 h	1000	1100	1200	1300	1400	1500 h and above
Frequency								
(a)	10	4	10	8	5	1	4	6
(b)	12	4	8	8	6	1	4	7
(c)	11	5	8	7	7	2	4	5
(d)	12	4	8	9	5	1	3	8

Qu. 4 The following table records long-jumps measured and rounded down to the nearest centimetre

Length of jump (metres)	No. of jumps
5.00–5.09	4
5.10–5.19	6
5.20–5.29	7
5.30–5.39	3
5.40–5.49	2
5.50–5.59	1

Which of the diagrams of Figure 16.10 accurately represents the histogram of the results?
Ans (a) figure A (b) figure B (c) figure C (d) figure D

Qu. 5 Figure 16.11 shows the cumulative frequency diagram (ogive curve) for the weight distribution of a number of tennis players. Determine the percentage of players with weights of 75 kilograms and above.
Ans (a) 66% (b) 75% (c) 25% (d) 34%

Qu. 6 and Qu. 7
Plot the cumulative frequency curve for following frequency distribution of incomes (class interval of £2000 is used):

Centre interval (£1000 units)	10	12	14	16	18	20	22
Frequency	1	5	9	12	7	4	1

Use the curve to determine:

Qu. 6 the median income (to nearest £200):
Ans (a) £16,000 (b) £14,000 (c) £15,300 (d) £14,700

Qu. 7 the lower quartile range:
Ans (a) £9,000 – £13,000
(b) 10,000 – £12,000
(c) £0 – £12,000
(d) £9,000 – £14,500

Qu. 8 Figure 16.12 is a histogram showing the frequency distribution of the marks recorded in an examination. Class intervals 0–9, 10–19, 20–29 ··· 80–89, 90–99 were selected. Distinction level was awarded to those students achieving a mark of 70 or better. How many students achieved the distinction grade?
Ans (a) 5 (b) 7 (c) 10 (d) 2

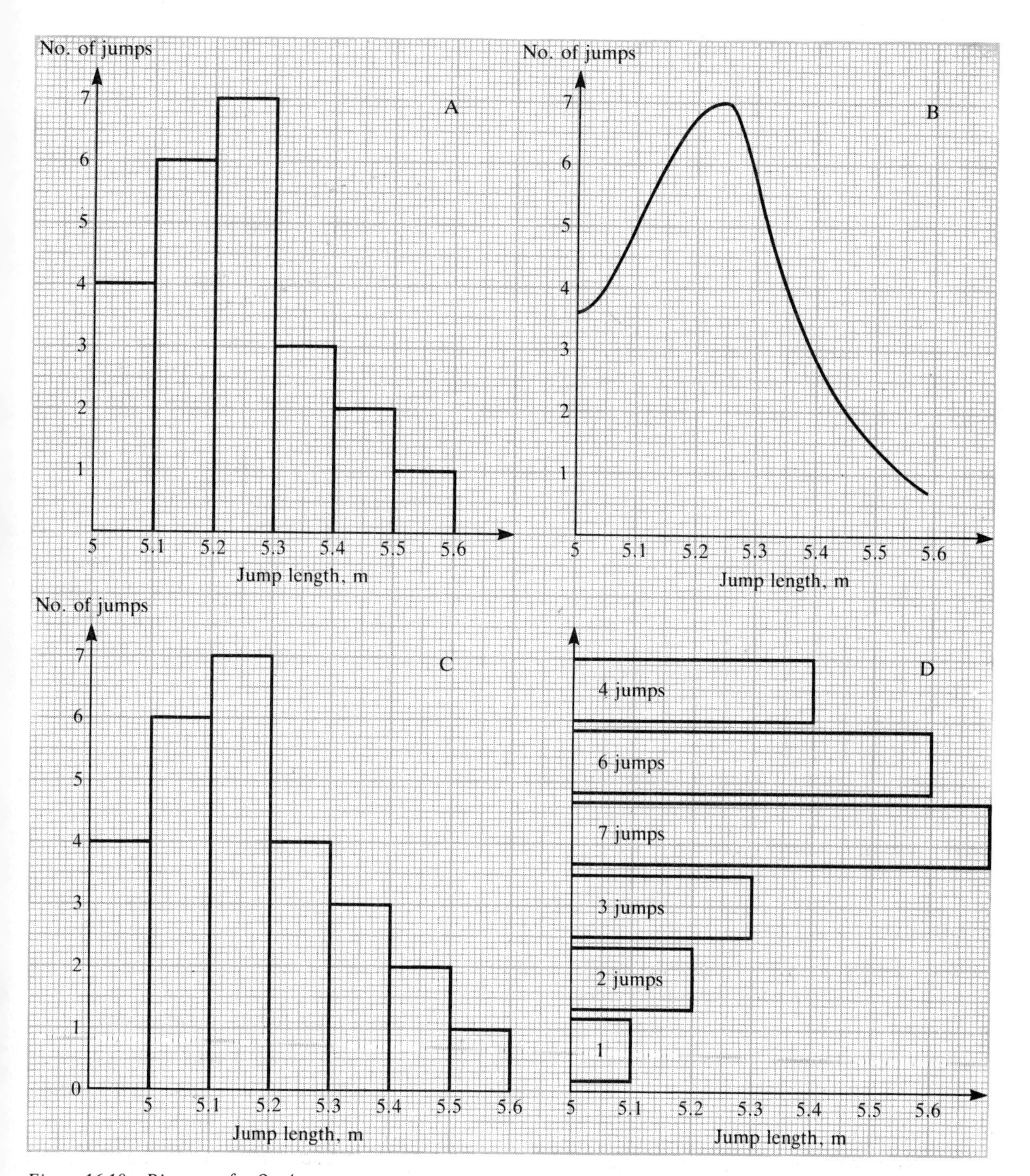

Figure 16.10 Diagrams for Qu 4.

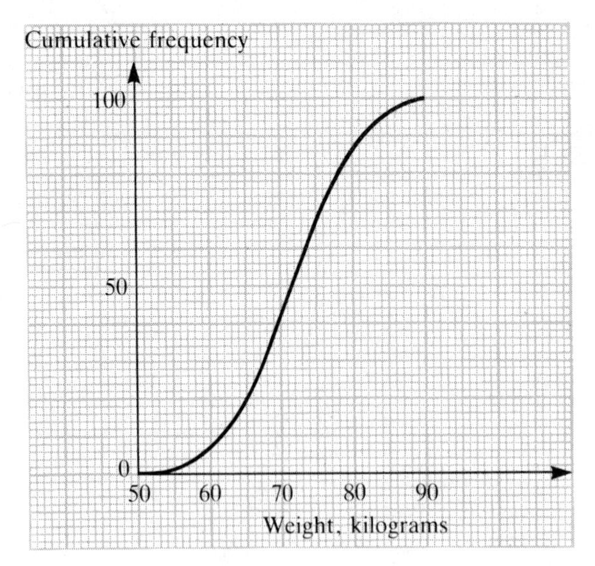

Figure 16.11 Cumulative frequency curve for Qu 5.

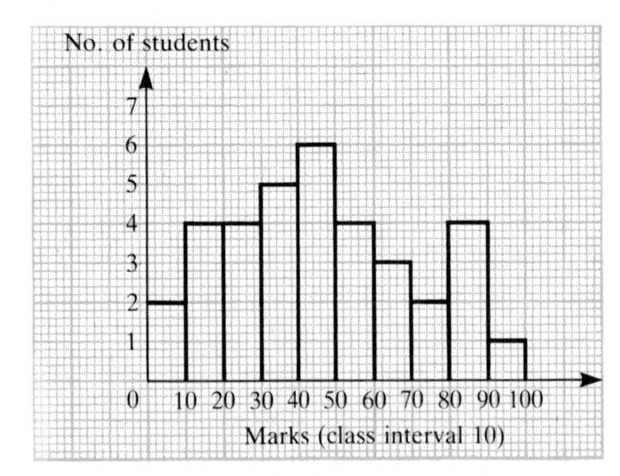

Figure 16.12 Histogram for Qu 8.

Problems 16

1 The following turnover of gross sales by regions for an international company was recorded for the year 1988–9:

Western Europe	£155.6 million
North Africa	£ 46.2 million
Asia	£122.8 million
South America	£ 18.6 million
North America	£247.0 million
United Kingdom	£493.2 million

Calculate the relative percentage turnover per region and represent the turnover in the regions by a 100% bar chart.

2 The following marks out of 100 were obtained by 30 candidates in an examination:

39	68	12	52	19	60
77	29	36	43	56	6
82	14	33	65	65	72
17	61	87	16	47	55
80	43	72	47	58	93

Use a tally diagram to group the marks into class intervals of 10 marks: 0–9, 10–19, 20–29 ··· 90–99 and determine the relative percentage frequencies of each class. Use these values to construct a pie chart to illustrate the frequency distribution of the results.

3 The annual trading figures over a five-year period for a company producing microprocessor circuits is depicted by the pictogram of Figure 16.13. Calculate:
(a) the percentage increase in trading from the 1983–4 year to the 1986–7 year;
(b) the percentage drop in trading in the 1987–8 year with respect to the 1986–7 figures.

4 A sample of 50 boxes of chocolates is selected at random from a batch of 10 000. The weight of each box is measured and rounded down to the nearest gram. The weights are given as:

Year Mc = £500,000
1983 – 84 Mc Mc
1984 – 85 Mc Mc Mc Mc
1985 – 86 Mc Mc Mc Mc Mc
1986 – 87 Mc Mc Mc Mc Mc
1987 – 88 Mc Mc Mc Mc

Figure 16.13 Pictogram for problem 3.

100	97	96	104	97	98	108	98	102	102
92	109	102	99	101	104	95	100	101	93
95	100	98	95	93	103	96	94	104	94
106	97	103	103	100	100	104	106	103	92
96	101	104	100	92	96	108	97	100	101

Use a tally diagram to group the box weights into classes having a class interval of 5 grams, starting with the class 90–94g. Determine the relative percentage frequency of each class and draw a histogram of the frequency distribution.

Estimate the likely number of boxes with weights within $\pm 5\%$ of 100g for the total batch of 10,000.

5 Figure 16.14 is a histogram showing the frequency distribution of the life-times of 1200 light bulbs. Using the histogram plot the cumulative frequency curve and use it to determine:
(a) the median life-time;
(b) the upper and lower quartile values;
(c) the semi-interquartile range.

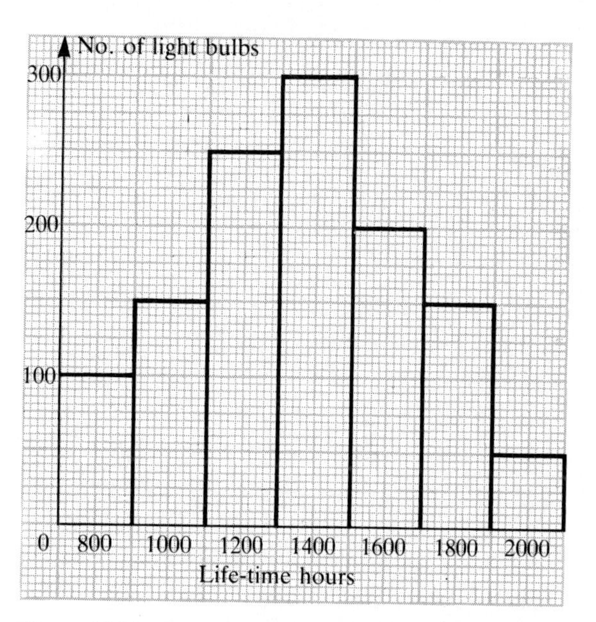

Figure 16.14 Histogram for problem 5.

Answers to Tests and Problems

Chapter 1: Indices and standard form

Test: MT1

Question No.	0	1	2	3	4	5	6	7	8
Answer	c	d	b	c	d	b	a	b	b

Problems 1

1 (a) 11^2 (b) 2^9 (c) 10^4 (d) 5^3

2 (a) $\frac{1}{4}$ or 0.25 (b) $\frac{1}{100}$ or 0.01 (c) 5
(d) $1\frac{1}{7}$ or 1.1429

3 (a) 16 (b) 16 (c) $\frac{1}{9}$ (d) 1.44

4 (a) 256 (b) $25 \times 32 = 800$ (c) $12 \times 2^5 = 384$

5 (a) 10^4 (b) 6 (c) 7^2 (d) 2^6

6 (a) 2.4×10^4 (b) 2.4×10^{-3} (c) 5×10^{-1}
(d) 9.999×10^2

7 (a) 8.37×10^6 (b) 3.6×10^{-3}
(c) 4.94×10^{-2} (d) 3.6×10^7
(e) 2.4×10^4 (f) 2.0×10^{-3}

Chapter 2: Indices and logarithms

Test: MT2

Question No.	0	1	2	3	4	5	6	7	8
Answer	b	c	a	c	c	d	a	d	d

Problems 2

1 (a) ± 3 (b) ± 12 (c) 6
(d) 125 (e) 25 (f) 10^5 or 100 000

2 (a) $\frac{1}{4}$ (b) $\frac{1}{16}$ (c) $\frac{1}{64}$
(d) $\pm\frac{1}{5}$ (e) $\pm\frac{1}{7}$ (f) $\pm\frac{1}{512}$

3 (a) 10^6 (b) 10^{-6} (c) $\pm 10^{-3}$ or $\pm\frac{1}{1000}$

4 (a) 3 (b) $5^2 = 25$ (c) $10^0 = 1$

5 (a) 2.0×10^{-4} (b) $3.69 \times 10^{-1} = 0.369$
(c) $7^{-1} = \frac{1}{7}$ (d) $9^\circ = 1$

6 (a) 4 (b) 4 (c) 3
(d) 2 (e) -2 (f) 1

7 (a) $2^6 = 64$ (b) $3^{-2} = \frac{1}{9}$
(c) 10^6 (one million)

8 (a) $\log 12 = \log 2 + \log 6 = 1.0791$
(b) $\log 3 = \log 6 - \log 2 = 0.4771$
(c) $\log 4 = \log 2 + \log 2 = 0.6020$

Chapter 3 Checking calculations and making approximations

Test: MT3

Question No.	0	1	2	3	4	5	6	7	8
Answer	c	a	c	c	b	a	b	c	c

Problems 3

1 (a) 36.79 (b) 0.07 (c) 488.00 (d) 0.00

2 (a) 64.54 (b) 579 700 (c) 0.000 477 9
(d) 2.718

3 (a) 32 (b) 16 (c) 5 (d) 0.6

4 (a) 11 (b) 4×10^8 (c) 36 (d) 2

5 (a) 300 mm (b) 1600 m (c) 1016 kg (d) 28g (e) 0.57 litre (f) 13 hp

6 (a) Not feasible, correct approximation 4×10^{-3}
(b) Correct
(c) Incorrect by factor of 10, correct approximation 200
(d) Correct to one significant figure, exact result £725
(e) Correct
(f) Incorrect by factor of 10, correct approximation 10 000.

Chapter 4: Using mathematical tables and charts

Test: MT4

Question No.	0	1	2	3	4	5	6	7	8
Answer	b	a	d	b	a	a	c	d	b

Problems 4

1 (a) 48.25 (b) 255,700 (c) 2.134×10^{-3} (d) 55.20×10^{-6}

2 (a) 5.126 (b) 0.8761 (c) 2.191×10^3 (d) 0.05167

3 (a) 0.3183 (b) 0.001764 (c) 7.042 (d) 8.333×10^5

4 (a) 0.798, 2.798, $\bar{3}.798$
(b) 3.6021, 6.6021, $\bar{6}.6021$
(c) $\bar{1}.8261$, $\bar{3}.8261$, $\bar{5}.8261$

5 (a) 4.708; 0.004 708; 47080
(b) 26.59; 0.026 59; 2 659 000

6 (a) 205.5 (b) 1.456 (c) 188.1
(d) 20.68 (e) 99.06 (f) 7.501
(g) 1.382 (h) 436.7 (i) 0.5245

7 (a) 0.18″ (b) 7.6 mm (c) 17.9 lb
(d) 2.5 kg (e) 81°C (f) −36°F

8 (a) $\frac{5}{36}$ (b) $\frac{5}{18}$ (c) $\frac{13}{18}$

Chapter 5: Using an electronic calculator

Test: MT5

Question No.	0	1	2	3	4	5	6	7	8
Answer	b	c	b	b	c	d	b	a	d

Problems 5

1 246.3
2 122.2
3 2502
4 270.8
5 231.8
6 32.15
7 0.029 55
8 1.530
9 0.3549
10 ±5.221
11 335.5
12 0.003 815
13 3981
14 0.2884
15 1.798

Chapter 6 Basic notation and rules of algebra

Test: MT6

Question No.	0	1	2	3	4	5	6	7	8
Answer	c	b	d	d	a	c	a	d	b

Problems 6

1 (a) 13 (b) −30 (c) 3

2 (a) 2 (b) $\frac{1}{2}$ (c) 192

3 (a) 1 (b) $3x^2 + 5xy + 11$
(c) $5x^3 + 3x^2 + 7x + 3$

4 (a) $8x^2 + 4x - 6$ (b) $x^3 + 9x^2 + 6x - 2$

5 (a) $144x^2y^2$ (b) $-27x^3y^3$ (c) $5xy^2$
(d) $x^3/8y^2$

6 (a) $7x-3$ (b) $4x^2-20x-12$ (c) $25a-35b$

Chapter 7 Multiplication and factorization of algebraic expressions

Test: MT 7

Question No.	0	1	2	3	4	5	6	7	8
Answer	c	c	a	b	b	d	b	d	a

Problems 7

1 (a) $4x^2+20x+25$ (b) $x^2+4x-21$
(c) $10a^2-7ab+b^2$

2 (a) $8x^2+4x$ or $4x(2x+1)$ (b) $2x^3+4xy+5xy^2$

3 $(x+1)(x+2)(x+3)=(x+1)(x^2+5x+6)$
$=x^3+5x^2+6x+x^2+5x+6$
$=x^3+6x^2+11x^4+6$

4 (a) $7x(4x+1)$ (b) $27(x-2y)$ (c) $9ab(4ab+1)$

5 (a) $(1-y)(1+y)$ (b) $(7x+5)(7x-5)$
(c) $25(2x+y)(2x-y)$

6 (a) $(3+a)(p+y)$ (b) $4(3t+1)(2-s)$

7 (a) $(3a+5b)(9a^2-15ab+25b^2)$
(b) $2(3x+1)(2-y)$

Chapter 8 The solution of simple and simultaneous equations

Test: MT 8

Question No.	0	1	2	3	4	5	6	7	8
Answer	c	d	b	b	a	c	d	c	a

Problems 8

1 (a) 16 (b) 4 (c) 5 (d) $3\frac{1}{2}$ (e) 30

2 (a) -2 (b) -1.97

3 $2\,\text{m/s}^2$

4 (a) $2\frac{3}{4}$ (b) 42

5 $\alpha=1.18\times10^{-5}$

6 (a) $x=-2, y=3$ (b) $x=1.5, y=2.3$

7 $a=0.5\,\text{m/s}^2, u=9\,\text{m/s}$

8 (a) $x=-3, y=5$ (b) $a=0.5, b=0.8$

9 $a=0.45, b=0.05$

Chapter 9 The evaluation and transformation of formulae

Multiple-choice test: MT 9

Question No.	0	1	2	3	4	5	6	7	8
Answer	c	b	d	c	a	c	b	a	b

Problems 9

1 (a) 400 (b) 9300

2 1.55

3 (a) [km] = 1.609 [miles]
(b) [pints] = 8 × 4.546 [litres]
(c) [kg] = 0.4536 [lb]; [lb] = 2.205 [kg]

4 (a) $(v-u)/t$ (b) $y-mx$ (c) $\sqrt[3]{(3v/4\pi)}$
(d) c/f (e) $\sqrt{(v/\pi h)}$ (f) $100I/(PT)$
(g) $1/(\omega^2 L)$ (h) $\sqrt{(Z^2-R^2)}$

5 (a) 11.73 ohms (b) 28 360 ohms

6 (a) 256 Hz (b) 0.349 76 metres

Chapter 10 Direct and inverse proportionality

Multiple-choice test: MT 10

Question No.	0	1	2	3	4	5	6
Answer	c	d	c	d	b	d	a

Problems 10

1 (a) $y=x$ (b) $y=3x$ (c) $y=-2x$

2 (a) $y=\frac{1}{2x}$ (b) $y=\frac{160}{x}$

3 (a) 0.524 (b) 33.52 (c) 4190

4 (a) 6 litres (b) 300 litres

5 (a) 1 N (b) 100 N

6 (a) 25 N (b) 1 N

Chapter 11 Equation of straight line graph

Multiple-choice test: MT 11

Question No.	0	1	2	3	4	5	6	7	8
Answer	c	c	b	a	b	d	c	b	b

Problems 11

1 (a) 3, 2 (b) 1, 0 (c) 2, −3
(d) 0,6 (e) −4,7 (f) no intercept, infinite gradient

2 (a) $\frac{1}{2}$ (b) $\frac{4}{3}=1.33$ (c) $1\frac{1}{2}$ or 1.5 (d) −2

3 (a) $y=3x+12$ (b) $y=-3x-12$
(c) $y=-3$ (d) $x=5$

4 (a) $y=x$ (b) $y=3x-6$ (c) $y=-3x+1$

5 (a) 47.5 (b) 7.9 (c) 6.8

6 (a) 140°F (b) 68°F (c) 82°C

7 (a) −14.2 (b) 2.2 (c) 4 (d) $x=-1$, $y=-7$

8 210×10^9

Chapter 12 Calculation of areas and volumes

Test: MT 12

Question No.	0	1	2	3	4	5	6	7	8
Answer	a	c	c	b	a	d	c	c	a

Problems 12

1 (a) $210\,m^2$ (b) $540\,m^2$ (c) $240\,cm^2$
(d) $63.275\,cm^2$

2 (a) $520\,cm^2$ (b) $193.05\,cm^2$

3 (a) $0.9426\,m^3$, $5.3414\,m^2$
(b) $2500\,mm^3$, $2530\,mm^2$
(c) $261.8\,cm^2$. $254.5\,cm^2$

4 (a) $2.3565\times10^{-5}\,m^3$, 0.181 45 kg
(b) $288\,m^3$, 371.52 kg

5 $5\times10^4\,cm^3$, $9562\,cm^2$

6 $735\,m^3$; $697.5\,m^3$, 6.975×10^5 kg, 4.1013×10^8 J, £1 139.25

Chapter 13 Types and properties of triangles

Test: MT 13

Question No.	0	1	2	3	4	5	6	7	8
Answer	d	d	a	b	a	c	b	b	d

Problems 13

1 (a) 74° (b) 50° (c) 68°

2 (a) 10.12 (b) 70.29

3 (a) congruent (SAS) (b) congruent (AAS)
(c) congruent (SAS) or (AAS) (d) congruent (RHS)

4 $x=4.86$, $y=8.24$

5 $a=9$, $b=7.2$, $c=5.4$

6 (a) 33 mm (b) AB = AC = 40 mm, isosceles
(c) 53°, 83°

Chapter 14 Geometric properties of circles

Multiple-choice test: MT 14

Question No.	0	1	2	3	4	5	6	7	8
Answer	c	b	d	b	b	c	d	c	d

Problems 14

1 (a) 157.1 mm (b) 134.0 m (c) 42 316 km

2 17.60 m

3 1669 km/h

4 (a) 2.095 m (b) 22.34 mm

5 2.546 m

6 (a) 65° (b) 124° (c) 32°
(d) 20° (e) $x = 110°$, $y = 105°$

Chapter 15 Introduction to trigonometry

Multiple-choice test: MT 15

Question No.	0	1	2	3	4	5	6	7	8
Answer	c	b	d	d	c	c	d	b	a

Problems 15

1 (a) $\sin\theta = 4/5$, $\cos\theta = 3/5$, $\tan\theta = 4/3$
(b) 12/13, 5/13, 12/5
(c) $1/\sqrt{2}$, $1/\sqrt{2}$, 1

2 (a) 0.7396 (b) 0.8028 (c) 5.1446
(d) 0.2571 (e) 1.0367 (f) 0.8454

3 (a) 74.66° (b) 65.85° (c) 58.28°

4 (a) $1/\sqrt{2}$ or $\sqrt{2}/2$, 1 (b) 1/2, $\sqrt{3}/2$, $1/\sqrt{3}$ (c) $\sqrt{3}$

5 (a) 0.4226 (b) 0.6249 (c) 0.2079

6 (a) $x = 30\cos 49° = 19.68$, $y = 30\sin 49° = 22.64$
(b) $h = 15.6/\sin 33° = 28.64$
(c) $a = 10/\sin 27° = 22.03$,
$b = 10/\tan 27° + 10/\tan 54° = 26.89$

7 61.93 m, 4335 m^2

8 28.87 m, 20.12 m

9 5.7 km north, 8.2 km east, 24.6 km from X

Chapter 16 Introduction to statistics

Multiple-choice test: MT 16

Question No.	0	1	2	3	4	5	6	7	8
Answer	c	b	d	b	a	d	d	a	b

Problems 16

1 W. Europe 14.36%, N. Africa 4.26%, Asia 11.33%,
S. America 1.72%, N. America 22.8%,
U.K. 45.52%

2 0–9: 3.3%; 10–19: 16.7%; 20–29: 3.3%,
30–39: 10% 40–49: 13.3%; 50–59: 13.3%;
60–69: 16.7%; 70–79: 10%; 80–89: 10%;
90–99: 3.3%

3 (a) 137.5% (rise = £2.375M − £1M)
(b) 10.5% (drop = £2.375 M − £2.125 M)

4 Relative % frequencies: 90–94g: 14%,
95–99g: 30%; 100–104g: 46%; 105–109 g: 10%.
7600 boxes

5 (a) 1260 hours approx.
(b) 1040 and 1480 hours approx.
(c) 220 hours approx.

Index